WATER TREATMENT MADE SIMPLE

For Operators

WATER TREATMENT MADE SIMPLE

For Operators

Darshan Singh Sarai, PhD

John Wiley & Sons, Inc.

Published by John Wiley & Sons, Inc., Hoboken, New Jersey
Published simultaneously in Canada

For general information about our other products and services, please contact our Customer Care Department within the United States at (800) 762-2974, outside the United States at (317) 572-3993 or fax (317) 572-4002.

Wiley also publishes its books in a variety of electronic formats. Some content that appears in print may not be available in electronic books. For more information about Wiley products, visit our web site at www.wiley.com.

Library of Congress Cataloging-in-Publication Data:
Sarai, Darshan Singh.
     Water treatment made simple for operators / Darshan Singh Sarai.
          p. cm.
     Includes bibliographical references and index.
     ISBN-13: 978-0-471-74002-5 (pbk.)
     ISBN-10: 0-471-74002-0 (pbk.)
     1. Water—Purification. I. Title.

     TD430.S24 2006
     628.1′62—dc22

2005005176

10 9 8 7 6 5 4 3 2

CONTENTS

PREFACE

Perhaps the most important person at a water treatment plant is the plant operator because that person is responsible for treating the water to meet or exceed the Federal Safe Drinking Water Act standards and public expectations. Operators need to be competent and confident to provide healthy water to protect the public health. However, when they are trying to determine the best way to accomplish this, operators do not often have the time or desire to wade through a sea of hard-to-understand technical texts written by engineers and university professors. Plant operators need a comprehensive and complete book written in a simplified and easy-to-understand manner. *Water Treatment Made Simple* should provide all the necessary tools to treat the water and resolve some of the day-to-day problems.

This book is the product of 30 years of experience in the water treatment field. During those years, I always felt a serious need for such a comprehensive and simplified book covering most of the aspects of water treatment and related subjects—such as hydraulics, applied math, chemistry, and microbiology—all in one volume. I envisioned it providing the basic knowledge to a student, giving a good start to a beginner in the water treatment field, being a refresher course for the people new in the field, serving as a ready reference for the supervisory staff, and preparing operators for the certification examination through review features.

For 20 years, I taught operating staff and treated water at the Water District # 1 of Johnson County, Kansas, from two polluted rivers, Missouri and Kansas, and the Kansas River Basin shallow wells. It was a unique opportunity and challenge to treat the water from these sources, which have a wide range of turbidity, pesticides, and a variety of other pollutants. The water is treated separately at two adjacent plants from two rivers, with the capability to combine the sources. Total capacity of the facility is about 200 million gallons per day. The classroom notes and detailed record of all the variations in the source water and day-to-day problems and their solutions are the main bases of this book.

To complete this work, I was fortunate to have the opportunity and responsibility to teach the operators, to set the guidelines for water treatment to exceed the Safe Drinking Water Act standards ahead of the compliance date, to handle and resolve the customer complaints, and to supervise the water quality control laboratory at the water treatment plant.

This book is intended to give the basic understanding to all persons interested in learning about the fundamentals of water treatment in an easy-to-understand format with illustrations. All the required subjects such as lab, hydraulics, math, chemistry, and microbiology are included for proper understanding of the treatment. Each chapter includes a brief and simplified version of fundamental principles involved in that phase of the treatment, expected problems and their possible solutions, and a set of self-study questions. Four appendixes are included for a quick reference. A list of references is given at the end of the book for further study. It is a complete, simplified book for the intended audience.

I sincerely acknowledge the encouragement and help of all the operators, colleagues, friends, and family members who have made this work possible. I am thankful, especially, to Bennet C. Kwan, retired superintendent of Treatment Plant, Water District #1 of Johnson County, Kansas, for his suggestions and reviewing the manuscript; Leslie Carreiro, lab technician, Water Treatment Plant, Asheville, North Carolina, for reviewing the manuscript; Colin Murcray, manager of Business Acquisitions for the American Water Works Association, for inspiring and helping me in many ways; A. P. Singh, president/owner of the Digital Printing Services, Kansas City, for helping in preparing and printing the draft; and James Harper, editor, John Wiley, and his staff (especially Lindsay Orman), for preparing and publishing the book.

I hope that the book will serve its intended purpose to help the operating staff to treat water effectively and confidently.

1

INTRODUCTION

Healthy water is vital for humans and for our progress. Our body contains 65 percent water by weight, which enables the body to perform various life functions. For example, it carries oxygen and digested food to different body parts, brings wastes like carbon dioxide and urea from those parts to the lungs and kidneys for their disposal, and participates in a number of biochemical reactions. An average adult drinks 2 liters (2.1 quarts) of water per day to carry out the vital functions; however, other personal needs require 20 gallons of water. A public water supply provides water for drinking, preparation of food, washing, sanitation, fire protection, swimming, and industrial use. Due to these requirements, U.S. water consumption averages 150 gallons per person per day. Treatment of raw water (from rivers, lakes, and wells) is required to provide safe drinking water—water without any harmful chemicals and without any waterborne pathogens (disease-causing microorganisms). Ideally, drinking water should be sparkling clear, cool, good tasting, reasonably soft, stable (neither corrosive nor scaling), plentiful, and cheap.

BACKGROUND

In ancient times, the Egyptians treated the drinking water by keeping it in large containers to settle out the sediments to make it look and taste better. Chinese boiled water to purify it. In Greece, Hippocrates, the father of medicine, around 400 b.c., found water as the carrier of waterborne diseases and suggested its boiling and cloth filtering to make it safe to drink. These ideas were the foundation for present-day sedimentation, disinfection, and filtration, which are the three major phases of water treatment. Currently, water treatment is becoming more sophisticated for an effective removal of pathogens

1

and harmful chemicals. Disinfection and filtration, in particular, are getting more attention. Disinfection is changing from plain chlorination to chlorine dioxide treatment, ozonation, or ultraviolet light treatment. Filtration is shifting from a commonly used high-rate sand filtration to membrane filtration.

A water treatment operator is responsible for treating the drinking water 24 hours a day to meet all the government requirements and the public expectations. The operator, therefore, needs a sound knowledge of fundamental concepts of water treatment and common problems with possible solutions during the treatment and related subjects, such as microbiology, chemistry, mathematics, and hydraulics. This book is intended to provide this information in a simplified manner. As a starting point, we will look at the primary regulation governing water quality—the Safe Drinking Water Act.

SAFE DRINKING WATER ACT

Although health agencies and the public were aware of the pollution of drinking water sources with contaminants such as waterborne pathogens and chemicals causing health problems, no serious action was taken until the 1970s. Until 1974, the drinking water quality in the United States was controlled by the U.S. Public Health Service (PHS), under water quality standards of 1962. These standards, originally developed in 1914, were revised twice in 1925 and again in 1962. Only 28 contaminants—including turbidity, coliform bacteria, lead, copper, and zinc—were regulated, and regulations were applicable only to the interstate water carriers. Other public water supplies used these regulations merely as guidelines for the drinking water quality.

Two reports, one by the PHS on Community Water Supply Study (CWSS) in 1970 and the other by the newly created U.S. Environmental Protection Agency (EPA) on the water quality study of the Mississippi River and New Orleans, Louisiana, in 1972, created serious concerns regarding the drinking water quality. The CWSS report showed that a large number of community water systems were not in compliance with the drinking water standards of 1962, and the EPA report showed the presence of a number of health-affecting synthetic organic compounds in the New Orleans drinking water supply. In 1974, the presence of cancer-causing trihalomethanes (THMs), formed by free-residual chlorine reacting with natural organic matter (NOM) in water, was also discovered.

Consequently, the U.S. Congress passed two major laws: the Clean Water Act (CWA) in 1972 and the Safe Drinking Water Act (SDWA) in 1974, the former to stop the pollution of water bodies and the latter to control the drinking water quality. The CWA requires wastewater treatment plants and industries to comply with the National Pollutant Discharge Elimination System (NPDES) to discharge the treated wastewater into the natural water bodies. The law was intended to improve, restore, and maintain the water quality of the water bodies for normal aquatic life and for recreational purposes, and assure a better quality of source water for drinking purposes.

The SDWA of 1974 (Public Law 93-523) regulated the drinking water quality with primary (enforceable) drinking water standards of various contaminants for all community water systems to protect the public health. Public water supply systems were designated as community water systems and noncommunity water systems. A *community water system* has 15 or more water connections, or 25 or more year-round customers. Noncommunity water systems were further classified as *nontransient noncommunity systems* if the same 25 or more people were served for at least six months of the year, such as schools, factories, and treatment plants; and *transient noncommunity systems* when people on transit were served (e.g., restaurants, gas stations, motels, and hotels). Water quality of noncommunity systems was classified as the responsibility of the supplier.

The SDWA of 1974 was amended in 1986 (also called the 1986 amendments of SDWA, as PL 99-339) and again in 1996 (also known as reauthorization bill, as PL 104-182). According to the SDWA, the federal regulatory agency is the EPA and the enforcement body is the state health department. Under the current SDWA act, the EPA regulates all the problematic physical, biological, and chemical contaminants in the drinking water. This process involves the state and the public for setting standards and monitoring the water quality. Each state must have its own water quality control program, with drinking water standards equal to or more stringent than those of the EPA. The public is kept informed through public notification in case of a serious problem with the drinking water supply and a yearly consumer confidence report (CCR). There are serious penalties for violations of National Primary Drinking Water Regulation (NPDWR) standards. Therefore, water treatment has become a very serious business.

Various contaminants—microbiological and chemical—are regulated by different rules under the SDWA:

1. *Microbiological quality.* Microbiological contaminants are controlled under the Surface Treatment Rule and the Coliform Rule. These rules emphasize the proper filtration and disinfection of all surface waters for adequate removal or inactivation of waterborne pathogens. Water quality is monitored daily for turbidity, proper disinfection, and the presence of coliform bacteria.

2. *Chemical quality.* Chemical quality is controlled by various rules, depending on the type and source of chemicals. There are five major groups of chemical contaminants:

    1. *Disinfectants and disinfection byproducts.* Contaminants such as trihalomethanes and haloacetic acids (HAAs) are controlled by the Disinfectants and Disinfection Byproducts Rule.

    2. *Corrosion byproducts.* Corroded metals such as lead and copper are controlled by the Lead and Copper Rule.

    3. *Agricultural chemicals.* Pesticides and other agricultural chemicals are controlled by the Synthetic Organic Compounds (SOC) Rule.

4. *Volatile organic compounds* (*VOC*). These contaminants are controlled through the Volatile Organic Compounds Rule.

5. Radionuclides. These are regulated under the Radionuclide Rule.

The EPA selects and regulates all the contaminants showing health effects (candidate contaminants) by setting the maximum contaminant levels (MCL), and maximum contaminant level goal (MCLG) under different rules. It revises them periodically. MCL should be as close to MCLG as feasible. MCLG is set at a level where a contaminant has no known adverse health effects with an adequate margin of safety. The benefits of MCL should justify the treatment cost. Every state has primary responsibility for enforcement of all the rules. The EPA has the authority to take action against public water systems in violation of the SDWA. Since 1998, all public water systems have been required to prepare and mail a yearly CCR to all customers in July showing all the contaminants detected during the year in the drinking water, their levels, and health risks associated with them. In case of a violation, a water utility is required by law to take appropriate action, inform the state agency, and notify the public. Thus, the water quality is assured to be adequate and safe for the public health.

Table 1-1 shows all currently regulated contaminants, their MCL, MCLG, health effects, and sources.

This is a brief overview of the SDWA. It is important for a utility to know the latest information on the SDWA because new regulations are promulgated, old regulations are revised, and more regulations are anticipated. For the latest information, contact the State Public Health Department, the American Water Works Association (AWWA), and the U.S. EPA. Their phone numbers and Web sites are given in the Appendix A.

Besides the compliance with the primary drinking water standards, the public expects the drinking water to be colorless, odorless, good tasting, nonstaining, and nondepositing type, for cosmetic and aesthetic reasons. For that, there are nonenforceable standards, which are called National Secondary Drinking Water Standards. Mostly, these are used as treatment guidelines for good public relations. Table 1-2 presents a list of 15 contaminants and their standards.

A plentiful and healthy water supply made civilizations flourish, and lack of it made them perish. A knowledgeable public water utility, especially operating staff, is essential for a community.

Table 1-1 U.S. EPA National Primary Drinking Water Contaminant Standards

Contaminant	MCLG[1] (mg/L)[2]	MCL or TT[1] (mg/L)[2]	Potential Health Effects from Ingestion of Water	Sources of Contaminant in Drinking Water
Microorganisms				
Cryptosporidium	zero	TT[3]	Gastrointestinal illness (e.g., diarrhea, vomiting, cramps)	Human and fecal animal waste
Giardia lamblia	zero	TT[3]	Gastrointestinal illness (e.g., diarrhea, vomiting, cramps)	Human and animal fecal waste
Heterotrophic plate count	n/a	TT[3]	HPC has no health effects; it is an analytic method used to measure the variety of bacteria that are common in water. The lower the concentration of bacteria in drinking water, the better maintained the water system is.	HPC measures a range of bacteria that are naturally present in the environment.
Legionella	zero	TT[3]	Legionnaire's Disease, a type of pneumonia	Found naturally in water; it multiplies in heating systems.
Total Coliforms (including fecal coliform and E. coli)	zero	5.0%[4]	Not a health threat in itself; it is used to indicate whether other potentially harmful bacteria may be present.[5]	Coliforms are naturally present in the environment, as well as feces; fecal coliforms and *E. coli* only come from human and animal fecal waste.

Table 1-1 (Continued)

Contaminant	MCLG[1] (mg/L)[2]	MCL or TT[1] (mg/L)[2]	Potential Health Effects from Ingestion of Water	Sources of Contaminant in Drinking Water
Turbidity	n/a	TT[3]	Turbidity is a measure of the cloudiness of water. It is used to indicate water quality and filtration effectiveness (e.g., whether disease-causing organisms are present). Higher turbidity levels are often associated with higher levels of disease-causing microorganisms such as viruses, parasites, and some bacteria. These organisms can cause symptoms such as nausea, cramps, diarrhea, and associated headaches.	Soil runoff
Viruses (enteric)	zero	TT[3]	Gastrointestinal illness (e.g., diarrhea, vomiting, cramps)	Human and animal fecal waste
Disinfection Byproducts				
Bromate	zero	0.010	Increased risk of cancer	Byproduct of drinking water disinfection
Chlorite	0.8	1.0	Anemia; infants & young children: nervous system effects	Byproduct of drinking water disinfection

Contaminant	MCLG	MCL	Potential Health Effects	Sources of Contaminant
Haloacetic acids (HAA5)	n/a[6]	0.060	Increased risk of cancer	Byproduct of drinking water disinfection
Total Trihalomethanes (TTHMs)	none[7] n/a[6]	0.10 0.080	Liver, kidney, or central nervous system problems; increased risk of cancer	Byproduct of drinking water disinfection
Disinfectants				
Chloramines (as Cl_2)	4[1]	4.0[1]	Eye/nose irritation; stomach discomfort, anemia	Water additive used to control microbes
Chlorine (as Cl_2)	4[1]	4.0[1]	Eye/nose irritation; stomach discomfort	Water additive used to control microbes
Chlorine dioxide (as ClO_2)	0.8[1]	0.8[1]	Anemia; infants & young children: nervous system effects	Water additive used to control microbes
Inorganic Chemicals				
Antimony	0.006	0.006	Increase in blood cholesterol; decrease in blood sugar	Discharge from petroleum refineries; fire retardants; ceramics; electronics; solder
Arsenic	0[7]	0.010 as of 01/23/06	Skin damage or problems with circulatory systems, and possible increased risk of getting cancer	Erosion of natural deposits; runoff from orchards, runoff from glass & electronics production wastes
Asbestos (fiber >10 micrometers)	7 million fibers per liter	7 MFL	Increased risk of developing benign intestinal polyps	Decay of asbestos cement in water mains; erosion of natural deposits

Table 1-1 *(Continued)*

Contaminant	MCLG[1] (mg/L)[2]	MCL or TT[1] (mg/L)[2]	Potential Health Effects from Ingestion of Water	Sources of Contaminant in Drinking Water
Barium	2	2	Increase in blood pressure	Discharge of drilling wastes; discharge from metal refineries; erosion of natural deposits
Beryllium	0.004	0.004	Intestinal lesions	Discharge from metal refineries and coal-burning factories; discharge from electrical, aerospace, and defense industries
Cadmium	0.005	0.005	Kidney damage	Corrosion of galvanized pipes; erosion of natural deposits; discharge from metal refineries; runoff from waste batteries and paints
Chromium (total)	0.1	0.1	Allergic dermatitis	Discharge from steel and pulp mills; erosion of natural deposits
Copper	1.3	TT[8]; Action level = 1.3	Short-term exposure: Gastrointestinal distress Long-term exposure: Liver or kidney damage People with Wilson's Disease should consult their personal doctor if the amount of copper in their water exceeds the action level	Corrosion of household plumbing systems; erosion of natural deposits

Contaminant			Potential Health Effects	Sources of Contaminant in Drinking Water
Cyanide (as free cyanide)	0.2	0.2	Nerve damage or thyroid problems	Discharge from steel/metal factories; discharge from plastic and fertilizer factories
Fluoride	4.0	4.0	Bone disease (pain and tenderness of the bones); Children may get mottled teeth	Water additive that promotes strong teeth; erosion of natural deposits; discharge from fertilizer and aluminum factories
Lead	zero	TT[8]; Action level = 0.015	Infants and children: Delays in physical or mental development; children could show slight deficits in attention span and learning abilities. Adults: Kidney problems; high blood pressure	Corrosion of household plumbing systems; erosion of natural deposits
Mercury (inorganic)	0.002	0.002	Kidney damage	Erosion of natural deposits; discharge from refineries and factories; runoff from landfills and croplands
Nitrate (measured as Nitrogen)	10	10	Infants below the age of six months who drink water containing nitrate in excess of the MCL could become seriously ill and, if untreated, may die. Symptoms include shortness of breath and blue-baby syndrome.	Runoff from fertilizer use; leaching from septic tanks, sewage; erosion of natural deposits

Table 1-1 *(Continued)*

Contaminant	MCLG[1] (mg/L)[2]	MCL or TT[1] (mg/L)[2]	Potential Health Effects from Ingestion of Water	Sources of Contaminant in Drinking Water
Nitrite (measured as Nitrogen)	1	1	Infants below the age of six months who drink water containing nitrite in excess of the MCL could become seriously ill and, if untreated, may die. Symptoms include shortness of breath and blue-baby syndrome.	Runoff from fertilizer use; leaching from septic tanks, sewage; erosion of natural deposits
Selenium	0.05	0.05	Hair or fingernail loss; numbness in fingers or toes; circulatory problems	Discharge from petroleum refineries; erosion of natural deposits; discharge from mines
Thallium	0.0005	0.002	Hair loss; changes in blood; kidney, intestine, or liver problems	Leaching from ore-processing sites; discharge from electronics, glass, and drug factories
Organic Chemicals				
Acrylamide	zero	TT[9]	Nervous system or blood problems; increased risk of cancer	Added to water during sewage/wastewater treatment
Alachlor	zero	0.002	Eye, liver, kidney, or spleen problems; anemia; increased risk of cancer	Runoff from herbicide used on row crops
Atrazine	0.003	0.003	Cardiovascular system or reproductive problems	Runoff from herbicide used on row crops

Contaminant			Potential Health Effects	Sources of Contaminant
Benzene	zero	0.005	Anemia; decrease in blood platelets; increased risk of cancer	Discharge from factories; leaching from gas storage tanks and landfills
Benzo(a)pyrene (PAHs)	zero	0.0002	Reproductive difficulties; increased risk of cancer	Leaching from linings of water storage tanks and distribution lines
Carbofuran	0.04	0.04	Problems with blood, nervous system, or reproductive system	Leaching of soil fumigant used on rice and alfalfa
Carbon tetrachloride	zero	0.005	Liver problems; increased risk of cancer	Discharge from chemical plants and other industrial activities
Chlordane	zero	0.002	Liver or nervous system problems; increased risk of cancer	Residue of banned termiticide
Chlorobenzene	0.1	0.1	Liver or kidney problems	Discharge from chemical and agricultural chemical factories
2,4-D	0.07	0.07	Kidney, liver, or adrenal gland problems	Runoff from herbicide used on row crops
Dalapon	0.2	0.2	Minor kidney changes	Runoff from herbicide used on rights of way
1,2-Dibromo-3-chloropropane (DBCP)	zero	0.0002	Reproductive difficulties; increased risk of cancer	Runoff/leaching from soil fumigant used on soybeans, cotton, pineapples, and orchards

Table 1-1 (Continued)

Contaminant	MCLG[1] (mg/L)[2]	MCL or TT[1] (mg/L)[2]	Potential Health Effects from Ingestion of Water	Sources of Contaminant in Drinking Water
o-Dichlorobenzene	0.6	0.6	Liver, kidney, or circulatory system problems	Discharge from industrial chemical factories
p-Dichlorobenzene	0.075	0.075	Anemia; liver, kidney or spleen damage; changes in blood	Discharge from industrial chemical factories
1,2-Dichloroethane	zero	0.005	Increased risk of cancer	Discharge from industrial chemical factories
1,1-Dichloroethylene	0.007	0.007	Liver problems	Discharge from industrial chemical factories
cis-1,2-Dichloroethylene	0.07	0.07	Liver problems	Discharge from industrial chemical factories
trans-1,2-Dichloroethylene	0.1	0.1	Liver problems	Discharge from industrial chemical factories
Dichloromethane	zero	0.005	Liver problems; increased risk of cancer	Discharge from drug and chemical factories
1,2-Dichloropropane	zero	0.005	Increased risk of cancer	Discharge from industrial chemical factories
Di(2-ethylhexyl) adipate	0.4	0.4	Weight loss, liver problems, or possible reproductive difficulties	Discharge from chemical factories
Di(2-ethylhexyl) phthalate	zero	0.006	Reproductive difficulties; liver problems; increased risk of cancer	Discharge from rubber and chemical factories
Dinoseb	0.007	0.007	Reproductive difficulties	Runoff from herbicide used on soybeans and vegetables

Contaminant	MCLG	MCL	Potential Health Effects	Sources of Contaminant
Dioxin (2,3,7,8-TCDD)	zero	0.00000003	Reproductive difficulties; increased risk of cancer	Emissions from waste incineration and other combustion; discharge from chemical factories
Diquat	0.02	0.02	Cataracts	Runoff from herbicide use
Endothall	0.1	0.1	Stomach and intestinal problems	Runoff from herbicide use
Endrin	0.002	0.002	Liver problems	Residue of banned insecticide
Epichlorohydrin	zero	TT[9]	Increased cancer risk, and over a long period of time, stomach problems	Discharge from industrial chemical factories; an impurity of some water treatment chemicals
Ethylbenzene	0.7	0.7	Liver or kidneys problems	Discharge from petroleum refineries
Ethylene dibromide	zero	0.00005	Problems with liver, stomach, reproductive system, or kidneys; increased risk of cancer	Discharge from petroleum refineries
Glyphosate	0.7	0.7	Kidney problems; reproductive difficulties	Runoff from herbicide use
Heptachlor	zero	0.0004	Liver damage; increased risk of cancer	Residue of banned termiticide
Heptachlor epoxide	zero	0.0002	Liver damage; increased risk of cancer	Breakdown of heptachlor
Hexachlorobenzene	zero	0.001	Liver or kidney problems; reproductive difficulties; increased risk of cancer	Discharge from metal refineries and agricultural chemical factories

Table 1-1 (Continued)

Contaminant	MCLG[1] (mg/L)[2]	MCL or TT[1] (mg/L)[2]	Potential Health Effects from Ingestion of Water	Sources of Contaminant in Drinking Water
Hexachlorocyclopentadiene	0.05	0.05	Kidney or stomach problems	Discharge from chemical factories
Lindane	0.0002	0.0002	Liver or kidney problems	Runoff/leaching from insecticide used on cattle, lumber, gardens
Methoxychlor	0.04	0.04	Reproductive difficulties	Runoff/leaching from insecticide used on fruits, vegetables, alfalfa, livestock
Oxamyl (Vydate)	0.2	0.2	Slight nervous system effects	Runoff/leaching from insecticide used on apples, potatoes, and tomatoes
Polychlorinated biphenyls (PCBs)	zero	0.0005	Skin changes; thymus gland problems; immune deficiencies; reproductive or nervous system difficulties; increased risk of cancer	Runoff from landfills; discharge of waste chemicals
Pentachlorophenol	zero	0.001	Liver or kidney problems; increased cancer risk	Discharge from wood-preserving factories
Picloram	0.5	0.5	Liver problems	Herbicide runoff
Simazine	0.004	0.004	Problems with blood	Herbicide runoff
Styrene	0.1	0.1	Liver, kidney, or circulatory system problems	Discharge from rubber and plastic factories; leaching from landfills
Tetrachloroethylene	zero	0.005	Liver problems; increased risk of cancer	Discharge from factories and dry cleaners

Contaminant			Potential health effects	Sources of contaminant in drinking water
Toluene	1	1	Nervous system, kidney, or liver problems	Discharge from petroleum factories
Toxaphene	zero	0.003	Kidney, liver, or thyroid problems; increased risk of cancer	Runoff/leaching from insecticide used on cotton and cattle
2,4,5-TP (Silvex)	0.05	0.05	Liver problems	Residue of banned herbicide
1,2,4-Trichlorobenzene	0.07	0.07	Changes in adrenal glands	Discharge from textile finishing factories
1,1,1-Trichloroethane	0.20	0.2	Liver, nervous system, or circulatory problems	Discharge from metal degreasing sites and other factories
1,1,2-Trichloroethane	0.003	0.005	Liver, kidney, or immune system problems	Discharge from industrial chemical factories
Trichloroethylene	zero	0.005	Liver problems; increased risk of cancer	Discharge from metal degreasing sites and other factories
Vinyl chloride	zero	0.002	Increased risk of cancer	Leaching from PVC pipes; discharge from plastic factories
Xylenes (total)	10	10	Nervous system damage	Discharge from petroleum factories; discharge from chemical factories

Table 1-1 *(Continued)*

Contaminant	MCLG[1] (mg/L)[2]	MCL or TT[1] (mg/L)[2]	Potential Health Effects from Ingestion of Water	Sources of Contaminant in Drinking Water
Radionuclides				
Alpha particles	none[7] ⎯⎯ zero	15 picocuries per liter (pCi/L)	Increased risk of cancer	Erosion of natural deposits of certain minerals that are radioactive and might emit a form of radiation known as alpha radiation
Beta particles and photon emitters	none[7] ⎯⎯ zero	4 millirems per year	Increased risk of cancer	Decay of natural and manmade deposits of certain minerals that are radioactive and might emit forms of radiation known as photons and beta radiation
Radium 226 and Radium 228 (combined)	none[7] ⎯⎯ zero	5 pCi/L	Increased risk of cancer	Erosion of natural deposits
Uranium	zero	30 ug/L as of 12/08/03	Increased risk of cancer, kidney toxicity	Erosion of natural deposits

Notes

[1] Definitions:

Maximum contaminant level (MCL). The highest level of a contaminant that is allowed in drinking water. MCLs are set as close to MCLGs as feasible using the best available treatment technology and taking cost into consideration. MCLs are enforceable standards.

Maximum contaminant level goal (MCLG). The level of a contaminant in drinking water below which there is no known or expected risk to health. MCLGs allow for a margin of safety and are nonenforceable public health goals.

Maximum residual disinfectant level (MRDL). The highest level of a disinfectant allowed in drinking water. There is convincing evidence that addition of a disinfectant is necessary for control of microbial contaminants.

Maximum residual disinfectant level goal (MRDLG). The level of a drinking water disinfectant below which there is no known or expected risk to health. MRDLGs do not reflect the benefits of the use of disinfectants to control microbial contaminants.

Treatment technique. A required process intended to reduce the level of a contaminant in drinking water.

[2] Units are in milligrams per liter (mg/L) unless otherwise noted. Milligrams per liter are equivalent to parts per million.

[3] EPA's surface water treatment rules require systems using surface water or ground water under the direct influence of surface water to (1) disinfect their water, and (2) filter their water or meet criteria for avoiding filtration so that the following contaminants are controlled at the following levels:

- *Cryptosporidium:* (as of January 1, 2002, for systems serving >10,000 and January 14, 2005, for systems serving <10,000) 99% removal.
- *Giardia lamblia:* 99.9% removal/inactivation.
- Viruses: 99.99% removal/inactivation.
- *Legionella:* No limit, but EPA believes that if *Giardia* and viruses are removed/inactivated, *Legionella* will also be controlled.
- Turbidity: At no time can turbidity (cloudiness of water) go above 5 nephelolometric turbidity units (NTU); systems that filter must ensure that the turbidity go no higher than 1 NTU (0.5 NTU for conventional or direct filtration) in at least 95% of the daily samples in any month. As of January 1, 2002, turbidity may never exceed 1 NTU, and must not exceed 0.3 NTU in 95% of daily samples in any month.
- HPC: No more than 500 bacterial colonies per milliliter.
- Long-Term 1 Enhanced Surface Water Treatment (effective date: January 14, 2005): Surface water systems or (GWUDI) systems serving fewer than 10,000 people must comply with the applicable Long-Term 1 Enhanced Surface Water Treatment Rule provisions (e.g., turbidity standards, individual filter monitoring, *Cryptosporidium* removal requirements, updated watershed control requirements for unfiltered systems).
- Filter backwash recycling: The Filter Backwash Recycling Rule requires systems that recycle to return specific recycle flows through all processes of the system's existing conventional or direct filtration system or at an alternate location approved by the state.

[4] More than 5.0% samples total coliform-positive in a month. (For water systems that collect fewer than 40 routine samples per month, no more than one sample can be total coliform-positive per month.) Every sample that has total coliform must be analyzed for either fecal coliforms or *E. coli* if two consecutive TC-positive samples, and one is also positive for *E. coli* fecal coliforms, system has an acute MCL violation.

[5] Fecal coliform and *E. coli* are bacteria whose presence indicates that the water may be contaminated with human or animal wastes. Disease-causing microbes (pathogens) in these wastes can cause diarrhea, cramps, nausea, headaches, or other symptoms. These pathogens may pose a special health risk for infants, young children, and people with severely compromised immune systems.

[6] Although there is no collective MCLG for this contaminant group, there are individual MCLGs for some of the individual contaminants:

- Trihalomethanes: bromodichloromethane (zero); bromoform (zero); dibromochloromethane (0.06 mg/L). Chloroform is regulated with this group but has no MCLG.
- Haloacetic acids: dichloroacetic acid (zero); trichloroacetic acid (0.3 mg/L). Monochloroacetic acid, bromoacetic acid, and dibromoacetic acid are regulated with this group but have no MCLGs.

[7] MCLGs were not established before the 1986 amendments to the Safe Drinking Water Act. Therefore, there is no MCL for this contaminant.

[8] Lead and copper are regulated by a treatment technique that requires systems to control the corrosiveness of their water. If more than 10% of tap water samples exceed the action level, water systems must take additional steps. For copper, the action level is 1.3 mg/L, and for lead is 0.015 mg/L.

[9] Each water system must certify, in writing, to the state (using third-party or manufacturer's certification) that when acrylamide and epichlorohydrin are used in drinking water systems, the combination (or product) of dose and monomer level does not exceed the levels specified, as follows:

- Acrylamide = 0.05% dosed at 1 mg/L (or equivalent)
- Epichlorohydrin = 0.01% dosed at 20 mg/L (or equivalent)

Table 1-2   National Secondary Drinking Water Standards

Contaminant	Secondary Standard
Aluminum	0.05 to 0.2 mg/L
Chloride	250 mg/L
Color	15 (color units)
Copper	1.0 mg/L
Corrosivity	noncorrosive
Fluoride	2.0 mg/L
Foaming agents	0.5 mg/L
Iron	0.3 mg/L
Manganese	0.05 mg/L
Odor	3 threshold odor number
pH	6.5–8.5
Silver	0.10 mg/L
Sulfate	250 mg/L
Total dissolved solids	500 mg/L
Zinc	5 mg/L

QUESTIONS

1. Why is water so important for life?

2. Give three important uses of water.

3. What is the role of Hippocrates in water treatment?

4. Explain the role of the U.S. EPA and a state in the Safe Drinking Water Act enforcement.

5. What happened in the 1970s to make the U.S. Congress act and pass water pollution and water quality control bills?

6. Define the terms *community water systems* and *noncommunity water systems*.

7. Explain the terms CCR, MCL, and MCLG.

8. In case of a violation of the SDWA, what is the course of action for the water utility?

2

WATER SOURCES

The total amount of water in the world is almost constant. It is estimated to be 3.7×10^{20} (370,000 quadrillion) gallons, and a very small amount (1.3×10^{12} gal) is added to it annually by burning gasoline. About 97 percent of this water is in the oceans, which is salty and unfit for human consumption without an expensive treatment. The remaining 3 percent is known as fresh water, out of which 2 percent is the glacier ice trapped at the North and South Poles. Thus, only 1 percent is available for drinking water. Some of the 1 percent is very deep and inaccessible subterranean (ground) water.

PROPERTIES

Pure water is a colorless, odorless, and tasteless liquid. The depth and light give it a blue or bluish-green tint. Tastes and odors in water are due to dissolved gases, such as sulfur dioxide and chlorine, and minerals. Water, a unique substance, exists in nature simultaneously as a solid (ice), liquid (water), and a gas (vapor). Its density is 1 g/ml or cubic centimeter. It freezes at 0°C and boils at 100°C. When frozen, water expands by one ninth of its original volume.

WATER CYCLE

Water is in continuous circulation through the water or *hydrologic cycle*, which is formed of three phases: atmospheric water, surface water, and groundwater.

Atmospheric Water

Atmospheric water is the moisture or water vapor in the air that forms clouds. The vapor is formed by the evaporation of water from water bodies and by transpiration from plants. *Transpiration* is the evaporation through stomata, small pores, on the underside of plant leaves. It is estimated that about 80,000 cubic miles of water from oceans and 15,000 cubic miles from lakes and the land surface evaporate annually. Only about 24,000 cubic miles of that falls back on the land surface; the rest falls back on oceans. The vapor is forced up by the warm air into the cooler levels, where it condenses to form clouds that create *precipitation* by the clumping of the vapor that falls back on the earth as rain, hail, snow, and sleet.

Surface Water

Surface water is the water on the earth's surface, such as oceans, lakes, rivers, and streams. Precipitation becomes a part of the surface water by falling directly into the water bodies and by runoffs.

Groundwater

Groundwater is the water deep in the ground. Some of the precipitation soaks through the soil and is called *infiltration,* a part of which is taken by plant roots. Another part of it moves back to the surface bodies, and a part moves deeper into the ground. This deeper movement of water through interstices or pores between sand particles is known as *percolation.* Percolated water is stopped by the impervious strata in the ground and becomes *groundwater,* which is retained in the sandy layers. The sandy layers saturated with water are known as *aquifers.*

Based on impervious strata, there are two types of aquifers: water table aquifer and artesian aquifer. *Water table aquifer* has only the bottom impervious layer, whereas an *artesian aquifer* has top and bottom impervious layers. Geologically, there are several types of soil forming the aquifers:

- *Gravel and sand* form the best aquifer.
- *Limestone and sandstones* are good water producers when sufficient voids and fissures are present.
- *Shale,* a hardened clay mixture, forms an aquifer if cracked.

Groundwater comes back to the ground surface through springs and wells. Surface water evaporates and becomes atmospheric water. Thus, the water cycle continues, resulting in use and reuse of the same water by various forms of life. Therefore, *precipitation is the only source of both surface and ground water supplies.*

Atmospheric water is contaminated only by particulate matter in the air and chemical vapor, such as automobile exhaust gases, and industrial flu gases. After the first part of rain, it is a good soft water, comparable to distilled water. Surface water is the most contaminated due to domestic and industrial discharges and runoffs. Groundwater due to natural filtration is generally the least contaminated. See Figure 2-1.

WATER SUPPLIES

There are two main water supplies: surface and ground.

Surface Water Supply

Surface water supply is the water from the lakes, reservoirs, rivers, and streams. These water bodies are formed of water from direct rain, runoffs, and springs. A *runoff* is the part of rain water that does not infiltrate the ground or evaporate. It flows by gravity into the water body from the surrounding land. This drainage area is known as the *watershed*. One inch of runoff rain/acre is equal to 27,000 gallons. The term *runoff coefficient* is the ratio of the runoff to rainfall, which varies with the topography and soil characteristics of the area. Watershed characteristics affect the water quality. Watershed protection is, therefore, important. *It is relatively easy and economical to control the contaminants at the source rather than removing them by the treatment.* The surface water treatment rule has provisions for an effective watershed control through sanitary survey, regular inspection, and protection.

Surface waters can be classified into *lentic* (lenis = calm), the calm waters, and *lotic* (lotus = washed), the running waters.

Lentic Water Supplies. Lentic waters are the natural lakes and impoundments or reservoirs (manmade lakes by building dams on running waters). Natural lakes of good quality water are very good sources of water. They need well-regulated sewage discharges to protect the water quality. Impoundments are useful, as they eliminate seasonal flow fluctuations and store water for adequate water supply, even under high consumer demand periods, such as drought in summer. Impounding also helps in the pretreatment of water by reducing turbidity by sedimentation and reducing coliform bacteria and waterborne pathogens through exposure to sunlight (ultraviolet part of sunlight is a disinfectant). Seepage and evaporation of water cause most of the loss of water from the impoundments. Algal growth and other planktons, drifters formed of free-floating algae, protozoans and rotifers, can cause taste and odor problems.

Life Stages of a Lake. Normally, a natural lake goes through an aging process called *eutrophication*. It starts with a beautiful young lake and ends as a fertile piece of land. This process in nature is very slow; it takes thousands of years for a lake to disappear. Humans, however, have accelerated this process by

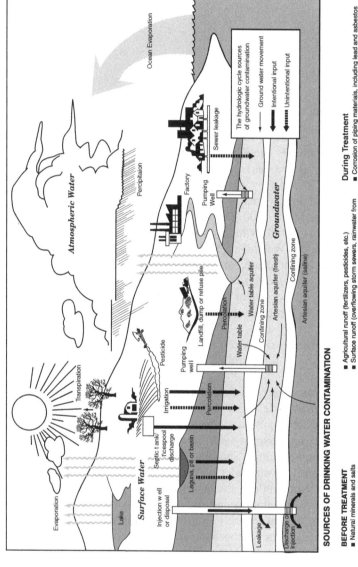

SOURCES OF DRINKING WATER CONTAMINATION

BEFORE TREATMENT
- Natural minerals and salts
- Decay products of radon, radium, and uranium
- Human and animal organic waste
- Defective storage tanks
- Leaking hazardous waste landfills, ponds, and pits
- Intrusion of salt water into depleted aquifer near the seashore
- Agricultural runoff (fertilizers, pesticides, etc.)
- Surface runoff (overflowing storm sewers, rainwater from oil-slicked or salt-treated highways, etc.)
- Underground injection of industrial waste

During Treatment
- Disinfection byproducts
- Other additives

During Treatment
- Corrosion of piping materials, including lead and asbestos
- Bacteria and dirt from leaking pipes
- Cross connections (incorrect pressure gradients that can suck polluted water into pipes instead of pushing it out)

Figure 2-1 Hydrologic Cycle and Sources of Water Contamination
(*Source: EPA Journal*, Vol. 12, No. 7, September 1986.)

adding nutrients and by discharging sewage, fertilizers, and detergents into lakes. There are three life stages of a lake: oligotrophic, mesotrophic, and eutrophic.

1. *Oligotrophic lakes* are young, deep, and clear, with few nutrients. They have a few types of organisms with low populations (e.g., Lake Superior).

2. *Mesotrophic lakes* are middle aged due to nutrients and sediments being continuously added. This results in more biomass productivity. With progressive pollution there is a great variety of organism species, with low populations at first. As time shifts and pollution increases, there will be higher populations of pollution-tolerant species. Sensitive species start disappearing. Sediments and dead organic matter make the lake shallower. At an advanced mesotrophic stage, a lake may have undesirable odors and colors in certain parts. Turbidity and bacterial densities increase. Lake Ontario is an example of this stage.

3. *Eutrophic lakes,* due to further addition of nutrients, have large algal blooms and become shallower, with fish types changing from sensitive to more pollution-tolerant ones. Biomass productivity becomes very high. Over a period of time, a lake becomes a swamp and finally a piece of land. Lake Erie is progressing toward this stage.

Reservoir Site. A reservoir site should be carefully selected by studying the following:

- Rate of stream flow
- Rainfall
- Type of soil and area of the watershed
- Water quality
- Soil type of the site
- Seepage rate

Protection of a reservoir should consider the following:

- No untreated sewage discharge
- Proper distance from recreational activities
- Possibly no railroad close by
- No grazing livestock within approved zones

Furthermore, a reservoir should have several intake openings to draw the adequate quantity and quality of water under different conditions.

Factors Affecting Lentic Water Quality. Several factors, such as temperature, sunlight, turbidity, dissolved gases, and nutrients, affect the water quality.

- *Temperature and stratification:* Water has maximum density (1 g/cm³) at 4°C. Above and below this temperature, water is lighter. Temperature changes in water cause *stratification* (layering) of water in lakes and reservoirs. Figure 2-2 illustrates the cycles of stratification over the course of a year. During summer, the top water becomes warmer than the bottom and forms two layers, with the top one warmer and lighter and the bottom one cooler and heavier. With the further rise in the temperature, a middle layer is formed, which is heavier than the top and lighter than the bottom layer. These three layers from top to bottom are known as *epilimnion, thermocline,* and *hypolimnion.* From the top to bottom, they are lightest and warmest, medium weight and warmer, and heaviest and coolest, respectively. There is no circulation of water in these three layers.

  If the water body is deep and the thermocline is below the range of the effective light penetration, oxygen supply is depleted in the hypolimnion because both photosynthesis and the surface source of oxygen are cut off. This condition is known as the *summer stagnation,* which causes the anaerobic decomposition of organic matter in the bottom sediments. During the fall, the temperature of the epilimnion drops until it is the same as that of the thermocline. Then the two layers mix. Finally, the temperature of the whole lake is the same, and there is a complete mixing. As the temperature of the top water reaches 4°C, it sinks to the bottom and bottom water moves to the top. This is known as *fall turnover.* This condition stirs the bottom mud and releases the anaerobic decomposition products such as sulfur dioxide and other odor-causing chemi-

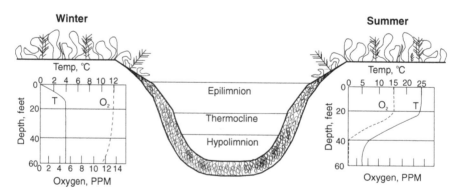

Figure 2-2 Stratification of a Lake

(*Source:* HDR Engineering, Inc. *Handbook of Public Water Systems, Second Edition.* Copyright © 2001 by John Wiley & Sons, Inc. Reprinted by permission of John Wiley & Sons, Inc.)

cals that cause severe taste and odor problems.

During winter, the epilimnion is at the lowest temperature; it could be covered with ice and is the lightest. Thermocline is at medium temperature and medium weight, while the hypolimnion is the heaviest, being about 4°C. This condition is called the *winter stratification*. In winter, due to lower biological activities, oxygen supply to the hypolimnion is less critical. Too much snow cover for longer time periods can cause oxygen depletion in the hypolimnion by reducing the light penetration— thus, the lower rate of photosynthesis. This condition can cause *winter fish kill*. In spring, as the ice of the epilimnion melts, there is mixing of the top two layers. As the temperature of the top layer reaches 4°C, water sinks once again to the bottom and results in the *spring turnover*, which, like fall turnover, can cause tastes and odors.

- *Light:* Light, the source of energy for photosynthesis, is important. The rate of photosynthesis depends on the light intensity and photoperiod (light hours per day). The amount of the biomass and oxygen production corresponds to the rate of photosynthesis. The amount of dissolved oxygen (DO) in the lakes is maximum at 2 P.M. and minimum at 2 A.M.

- *Water movements:* Water movements, such as wave action in the lakes, mix the dissolved oxygen at the interphase of air and water into deeper layers that increase the rate of absorption of oxygen from the atmosphere.

- *Turbidity:* Turbidity affects the rate of the penetration of sunlight, and thus, photosynthesis. The more the turbidity, the less is the sunlight, and vice versa. The less the sunlight, the lower is the rate of photosynthesis, and consequently less the DO.

- *Dissolved gases:* These are mainly carbon dioxide (CO_2) and oxygen (O_2). Carbon dioxide is produced during respiration and is used in photosynthesis; oxygen is produced during photosynthesis and is needed for respiration. DO is consumed by the microorganisms for the aerobic decomposition of biodegradable organic matter. This oxygen demand of the water is known as *biochemical oxygen demand (BOD)*. The more the BOD, the less is the DO in the water. The more the DO, the better is the quality of water, and vice versa. The minimum amount of DO to maintain normal aquatic life, such as fish, is 5 mg/L.

- *Nutrients:* These are both organic and inorganic chemicals. They enter the water body through point and nonpoint discharges. A *point discharge* is an entry at a designated point such as a sewage discharge, and a *nonpoint discharge* is a discharge at any undefined point such as farmland and feed lot runoffs.

The organic nutrients are formed of natural organic matter (NOM) and organic matter contributed by man through sewage and industrial wastes. All of the organic matter is represented as total organic carbon (TOC). NOM includes humic acid and fulvic acid, the precursors of the trihalomethanes

(THMs), which can cause cancer. Too much organic matter can cause high BOD that can result in anaerobic conditions. Thus, the less organic matter in the water, the better is the water quality. The old notion, *"Dilution is the solution to water pollution,"* is not true anymore due to higher population and modern civilization. The dilution factor is not effective. Inorganic nutrients are biogenic salts, such as chlorides, sulfates, nitrates, and phosphates, which are contributed by farmland runoffs and by dissolving the natural minerals from the soil. They can allow the excessive growth of algae known as *algal blooms.*

Lotic Water Supplies. Rivers, streams, and springs are lotic water supplies.

Life Stages of a River. The life stages of a river or a stream can be divided into four stages: establishment or birth, young stream, mature stream, and old stream.

1. *Establishment or birth* of a stream may be a seasonal run or a headwater from a spring or a lake.
2. *A young stream* is a stream that has become a permanent stream with the bed eroded below the water table and thus receives ground water. A stream at this stage is very clean with very few organisms. It is termed *pristine.*
3. *Mature stream* is the stage when a stream has traveled down to the plains. It has become wider, deeper, and turbid with low velocities. The water is generally warmer and the bottom has sand, mud, silt, or clay with organic matter. Water quality has deteriorated, and the floodplain and valleys are developed.
4. *Old stream* is a stage when a stream has reached the geologic base level. The floodplain may be broad and flat. During the normal flow periods, the channel is refilled and many shifting bars are formed. The velocity is low and water quality may be very poor.

Factors Affecting Lotic Water Supplies. As compared to lakes and reservoirs, running water is affected mainly by current and nutrients.

- *Current:* It is the velocity or rate of flow of water. The faster the current, the better it is. Current mixes the oxygen from the atmosphere and keeps the bottom of the stream clean by washing away the settlable solids. There is more DO and less natural organic matter (NOM) that would otherwise decompose in the bottom. Thus, due to the current, streams and rivers seldom go anaerobic.
- *Nutrients:* Main sources of nutrients are drainage from the water shed, point and nonpoint discharges. There is more sewage and nonpoint sources discharge into the rivers than into the reservoirs due to the length of the river. Heavy rains and drought conditions also cause serious prob-

lems, such as high turbidity and more nutrients. Rivers and streams are supposed to be recovered from the effects of normal domestic sewage discharge within 20 miles. Organic matter from sewage was supposed to be only a little extra food for aquatic life like bugs and critters. However, it is not true anymore.

Surface water supply is the most contaminated supply, mainly due to discharge of sewage, used water, which is the source of waterborne pathogens, runoffs from farmland, which are the source of *Cryptosporidium,* pesticides, and fertilizers; and industrial discharges, which are the source of a variety of contaminants. Surface water, therefore, needs the maximum treatment for potability.

Groundwater Supply

Underground water is supposed to be the purest form of natural water. Sometimes, it is so pure that it does not need any further treatment for drinking purposes. It is the least contaminated and has very low turbidity due to natural filtration of the rain water. Generally, it is cool, clear, and odorless. It can be contaminated by underground streams in areas with limestone deposits, septic tanks discharge, and underground deep well leaks. Therefore, it may need disinfection. It needs only mineral removal treatment when compared to surface water supply. It contains more dissolved minerals such as calcium, magnesium, iron, manganese, and sulfur compounds than the surface supply. There are two sources of groundwater: springs and wells.

Springs. Whenever an aquifer or an underground channel reaches the ground surface such as a valley or a side of a cliff, water starts flowing naturally. This natural flow is known as a *spring*. A spring may form a lake (at the bottom of a valley), a creek, or even a river. The quantity and velocity of a spring flow depend on the aquifer size and the position of the spring relative to the highest level of the water table. The larger the aquifer, the more is the flow; the lower the spring site (than the water table), the higher is the velocity, and vice versa. Regions with limestone deposits have large springs as the water flows in undergound channels, formed by the erosion of limestone.

Some springs have a special quality of water; some are believed to have medicinal value due to certain minerals in their water, and some have very warm water. The quality of the spring water depnds on the nature of the soil through which the water flows. For example, a mineral spring has dissolved minerals, a sulfur spring has dissolved sulfur, and a hot spring has hot water as the water flows through volcanic rocks.

Wells. Public groundwater supply is usually well water because springs are rare. A well is a device to draw the water from the aquifer. Deeper wells (more than 100 feet) have less turbidity, more dissolved minerals, and less bacterial count than shallow wells. Shallow wells have less natural filtration of water due to less depth of the soil. If the turbidity and bacterial counts of the well water fluctuate with the rainfall, then this water is considered under the direct influence of surface water. According to the SDWA, water from such wells needs to be treated just like the surface water. Well water may need only some disinfection; commonly, free residual chlorine does the job. Generally, public perception is that the treated water should be comparable to the groundwater. Small rural communities (less than 25,000 populations) generally use the groundwater from wells. About 35 % of the American population uses groundwater supply.

Well Protection. Protection of well water is important. First, select the proper site by drilling test holes and keeping in mind various sources of well contamination. For proper sanitary protection, specifications are devised by the engineering division of the state health department. Here are some guidelines:

- *Distance from human wastes.* A well should be properly located in regard to sewer lines and septic tank drain fields. The distance should be 50 feet. Sewage treatment plant, animal feed lots, sanitary landfill, and habitation should all be 500 feet away from a well. These are some generally recommended distances. Due to geological variations, each state has its own guideline for the well location.
- *Protection from geological leakage.* Only water table wells need good protection. This can be accomplished by having a two-inch-thick cement layer around the well pipe.

Classification of Wells. Based on the type of construction, wells are grouped as drilled, bored, dug, or driven.

- *Drilled wells.* They are deep wells used by the water utilities. They are of two types: *water table* and *artesian* wells, depending on the type of the aquifer. Water table wells use the water table aquifer, whereas artesian wells use an artesian aquifer. The water level in an artesian well can rise above the level of aquifer, depending on the location of the well relative to the highest level of the aquifer. If a well site is 20 feet below the highest level of aquifer, the water will rise in the well to 20 feet. without pumping to seek the highest level of the aquifer. If the well site is above the highest level of the aquifer, then water will not flow, and pumping is required.

    For the public water supply, wells are 8 to 24 inches in diameter and normally 400 to 1,000 feet apart. A well is generally cased with iron or steel pipe from the ground surface to the aquifer. For the proper protection, casing must pass through one impervious soil layer, must be tight at joints, and must be free from any corrosion leakage. A well pumping house should be above flood-water level. The space between the casing and the drill hole is filled with cement poured down to the impervious stratum for adequate seal to prevent any infiltration. The water is admitted from the aquifer to the well through a brass strainer at the lower end of the pipe. Generally, the area outside the strainer is packed with gravel for easy flow of water (see Figure 2-3).

- *Bored wells.* They are excavated with hands or power augers. Usually they are quite shallow.
- *Dug wells.* They are 5 to 40 feet in diameter. They are hard to case. The casing is the pipe or the wall retaining the water. Casing should be at least below the aquifer level and a foot above the ground level.

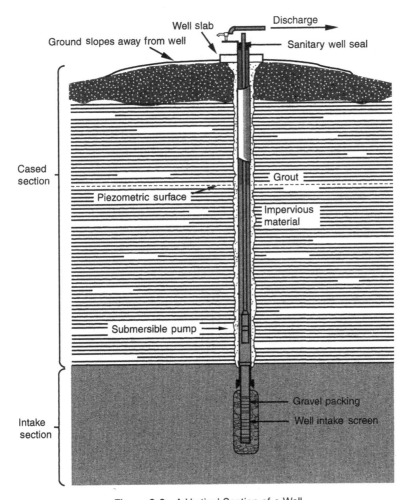

Figure 2-3 A Vertical Section of a Well

(*Source: Water Treatment Operator Handbook.* American Water Works Association, 2002.)

- *Driven wells.* They are constructed by driving a pipe, with a pointed screen attached to its end, into the aquifer. They are shallow (not more than 30 feet deep) with small capacity and suitable for areas with relatively soft soil.

Developing a Well. A well is developed by pumping out the mud and washing the well, until the discharge water is clear, as shown in Figure 2-4. The height to which water rises when not pumping is called the *static level.* It is the distance between the water level and the ground surface. During pumping, the water level drops to a point called the *pumping level.* The difference between the static level and the pumping level is called the *drawdown.* The capacity of the well per foot draw down is called the *specific capacity.* If a well capacity is 200 gallons per minute (gpm) and its drawdown is 10 feet, specific capacity of the well is 200 gpm/10 ft, or 20 gpm/ft. The radius of the circular area of aquifer, which is dewatered around the well while pumping, is known as *radius of influence.* This dewatered area becomes a depression like an inverted cone called the *cone of depression.* If the water level does not reach the original static level when pumping is stopped, the distance it falls short is called the *residual drawdown.*

Factors Affecting the Well Yield. Well yield or capacity is the rate of water production by the well expressed as gallons per minute.

- *Draw down.* The yield is proportional to the draw down. The more the draw down, the greater the yield.
- *Pump diameter.* Doubling the diameter size of the well pipe will increase the yield by about 15 percent.

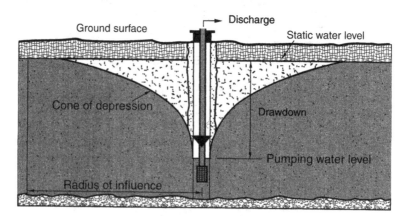

Figure 2-4   A Well While Pumping

(*Source: Water Treatment Operator Handbook.* American Water Works Association, 2002.)

- *Aquifer.* The yield is directly proportional to the depth of the well in the aquifer. If a well penetrates 20 feet into the aquifer, the yield will be doubled if the penetration depth is increased to 40 feet with the same drawdown.
- *Nature of the sand particles.* The coarser the particles, the easier the movement of the water, and the higher is the yield.
- *Distance from another well.* The more the distance, the better it is. There should not be any overlapping of the areas of the influence of different wells.

Stimulation of a Less Productive Well. Common causes of lower yield of a well are overpumping, clogging of the screen, or aquifer depletion. Here are some commonly used corrective measures:

- *Surging.* This is alternately pumping and backwashing to force the clogging material out. It is also done by forcing compressed air against the screen or by using a *surge block* up and down in the well casing.
- *Pumping at a higher rate.* The higher rate of pumping will clear the sand from the screen.
- *Chlorination.* Chlorine treatment of the wells has been successful for removing bacterial slimes and dissolving calcium carbonate ($CaCO_3$) deposits. Several hours' contact and repeat treatment are required.
- *Sodium hexametaphosphate and chlorine treatment.* For wells with high hardness and high iron, a mixture of 15 to 30 pounds of sodium hexametaphosphate and 1 to 2 pounds of bleach/100 gallons of solution is poured down the well and allowed to stay for one to two days. This solution kills the iron bacteria and dissolves rust and calcium carbonate buildup on the screen and improves the well capacity. After treatment, wash the well until the water quality is acceptable.

Water is the most abundant, most used, and most abused natural resource. We need to be water wise to enjoy it ourselves, and to allow our children and their children to enjoy it.

QUESTIONS

1. Write three parts of the water cycle and explain how they form the cycle.

2. Define these terms:
   a. Transpiration
   b. Infiltration
   c. Percolation

    d. Aquifer

    e. Water table

3. State three main advantages of an impoundment water supply as compared to a river.

4. What causes the stratification of a lake?

5. Name three layers of a stratified lake.

6. Explain these terms:

    a. Hypolimnion

    b. Mesotrophic

    c. Eutrophic

    d. BOD

7. Normally, a river does not stratify or go anaerobic. Explain why.

8. a. What percentage of the U.S. population uses groundwater?

    b. Why is groundwater considered a better source of drinking water than other sources?

9. a. Discuss various parts of a well.

    b. How does a water table well differ from an artesian well?

    c. Explain these terms:

        a. Static level

        b. Pumping level

        c. Cone of influence

        d. Specific capacity

10. State three main factors affecting the yield of a well.

11. About 80 percent of the U.S. population uses surface water supply. T or F

12. Write three sources of human wastes that can contaminate a well if it is not properly located.

3

WATER TREATMENT LABORATORY

A laboratory is a place for precise work to determine appropriate treatment of raw water and the quality of the finished water. It must be kept properly organized, well maintained, and scrupulously clean. All instruments must be kept clean and routinely calibrated with proper records. A number of lab tests are needed daily, quarterly, semiannually, annually, and at other specified intervals to monitor the water quality before, during, and after the treatment. A test is not better than the sample, and the sample is not better than the manner in which it is collected.

SAMPLING

Valid testing starts with an adequate and representative sampling. A sample is either a grab or a composite. A *grab sample,* as the name indicates, is a specific volume collected at one site at one time. These samples indicate the quality of water at that time and at that site. Grab samples are taken for bacteriological and disinfection residual tests. A *composite sample* is a mixture of a number of portions taken at the specific intervals (e.g., samples for mineral analysis because minerals in water are more stable). This reduces the number of tests. Each portion can be proportionate to the flow or volume.

For each test operators should follow the prescribed sampling size, collecting, and preserving procedure given in the Standard Methods for the Examination of Water and Waste Water (Standard Methods). Testing must be done as soon as possible and not later than the specified holding time.

QUALITY ASSURANCE OR QUALITY CONTROL

To assure the validity of the results, a water quality control laboratory requires a good quality-assurance program. For certification, this program has both internal and external quality-control measures. Internal quality control is like running a standard with the sample, testing duplicate samples, and having standard curves. External quality control has laboratory certification by the state for all parameters performed by the laboratory. It requires successfully analyzing the performance samples twice a year. Furthermore, there is an onsite visit by the State Certification Personnel to check the equipment, procedures, laboratory personnel, and records.

TESTS

Various regularly performed common tests by the operating staff are for tastes and odors, turbidity, jar test, pH, alkalinity, hardness, disinfection residual, coliform bacteria, and the heterotrophic plate count. All other tests are run either by highly trained chemists and microbiologists of the lab or by certified contract laboratories.

Tastes and Odors

Testing for taste and odor is important because of aesthetic value. The majority of water quality complaints are of this type. Most of the organic and some inorganic chemicals cause tastes and odors. These chemicals come from the decaying organic matter, runoffs, industrial wastes, and municipal sewage discharges. Geosmin and methyl-isobarneol (MIB) are the serious odor-causing chemicals; they are produced by bacteria—particularly actinomycetes—while decomposing dead organic matter at the bottom of the water bodies. Even a very low concentration of these chemicals can cause earthy-musty odors. These odors are common in spring and fall due to the turn over of the lakes and reservoirs. In the groundwater, the tastes and odors can be due to iron, manganese, and hydrogen sulfide (H_2S).

These are general classes of odors:

- Aromatic (spicy)
- Balsamic (flowery)
- Chemical
- Disagreeable
- Earthy
- Musty
- Grassy
- Vegetable

These are called the *reference odor* in the water samples.

Principle. The sample is diluted with odor-free water until a dilution is found with the least detectable odor; this dilution factor is the *threshold odor number.* The sense of smell varies widely—even in the same individual and from person to person. Therefore, it is recommended that a panel of individuals with keen sense of smell be used for an accurate determination of the odors in the water. Water samples are heated in a water bath at 60°C to vaporize odor-causing chemicals. These odors are sniffed after sniffing the odor-free water.

Procedure. Add 200 mL, 100 mL, 50 mL, 12 mL, and 2.8 mL portions of the representative sample sequentially to (thoroughly cleaned and rinsed with odor-free water) correspondingly marked 500 mL Erlenmeyer flasks and dilute each portion to 200 mL with odor-free water. Apply a glass stopper to each of them and heat in a water bath for 30 minutes at 60°C. Agitate the flask with 0 mL sample (control or odor free), remove the stopper, and sniff the vapors. There should not be any odors in the control sample. Repeat this with the flask containing the undiluted water. If it has an odor, this is the *reference odor.* Repeat the sniffing procedure by first sniffing odor-free water and then the flask with the lowest volume (2.8 mL) of the sample. Continue progressively to the highest sample volume until the dilution with the detectable odor is determined. Dilution factor of the dilution with the detectable odor is the threshold odor number. For example, we are testing raw river water with musty-earthy odor and 50 mL sample is the dilution with some musty odor. Threshold odor number of this sample is 200mL/50mL or 4, and the reference odor is musty-earthy. This means that people can easily detect these odors in the water.

Turbidity

Turbidity is the murkiness in the water caused by colloidal (1 to 100 nanometer (nm) particles) and other suspended particles, such as clay, sand, silt, organic matter of plant and animal origin, planktons, and other microscopic organisms. Turbidity particles can be waterborne pathogens or particles harboring them. The lower the turbidity, the less is the amount of the particulate matter. It means there is less probability of the presence of waterborne pathogens, and the water is safer. Therefore, turbidity is one of the primary standards for the drinking water. The finished water turbidity is tested at least every four hours. Turbidity of the finished water should be equal to or less than 0.3 nephalometric turbidity unit (NTU) in 95 percent of the samples/month.

Principle. Turbidity is measured as the amount of scattered light by the suspended particles in the sample. The turbidity unit, NTU, is based on the amount of light scattered by particles of formazine, a polymer, used as a reference standard due to the reproducibility of the results. One mg/L of formazine equals 1 NTU.

Procedure. Collect a representative grab sample below the surface of water, mix it, and pour into a clean and scratch-free measuring cell. Pour the sample slowly to avoid air bubble formation. To avoid any smudges from the hands, hold the cell from the top part. Scratches, smudges, and air bubbles give false higher readings.

Calibrate the turbidity meter and determine the turbidity. Follow the procedure according to the manufacturer's instructions.

Jar Testing

Jar testing is a useful tool to determine the practical optimum dose of a chemical under the simulated plant conditions. It uses a range of increasing dose of a particular chemical in a series of six jars with a stirring and illumination mechanism (see Figure 3-1). Most of the problems in the source water (particularly in the surface water) quality are due to seasonal variations or other unusual circumstances, such as drought, heavy rains, unexpected discharge of raw sewage, or runoffs from farm land. These problems can be solved by this test, which is important for coagulation, softening, sedimen-

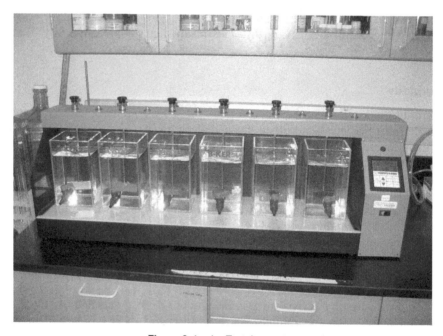

Figure 3-1 Jar Test Apparatus

(*Source:* MWH, *Water Treatment, 2e.* Copyright © 2005 by John Wiley & Sons, Inc. Reprinted by permission of John Wiley & Sons, Inc.)

tation, removal of synthetic organics (such as atrazine), and for tastes and odor control. It makes the water treatment more effective, easy, and economical.

Prepare the required dosing solutions and have the jars clean and washed.

Dosing Solutions. Weigh the chemical as accurately as required to make a dosing solution: 1 g/L of the chemical equals 1 mg in 1 mL of the dosing solution; 5 g/ L, 5 mg in 1 mL, and so on. When we apply 1 mL of the dosing solution to a liter of sample, it gives the corresponding number of mg/L of the chemical (e.g., 1 mL of the 5 g/L of dosing solution gives 5 mg/L dose when applied to 1 liter of water). For rough estimates, 1 mL equals 20 drops; thus, 2 drops mean 0.1 mL of the dosing solution. Sometimes drops are used for lower doses.

General Procedure. Simulate the treatment plant conditions for each jar test and pour 1 liter of representative water sample into each of six jars and apply a series of six doses of the required chemical, starting with a control (0 dose) and ending with the highest dose. Mix the chemical and the sample by using the mixing paddles. Treat it under the plant conditions, and check the results.

Example. Determine the polymer dose to treat 100 NTU turbidity in the raw water sample. Suppose an estimated optimum dose of polymer for effective removal of turbidity is 8 mg/L. Dose the jars of the water sample with 0, 4, 6, 8, 10, and 12 mg/L polymer, respectively. Immediately, lower the stirring paddles and stir at 60 revolutions/minute (RPM) for 0.5 minutes for flash mixing. Then reduce the stirring speed to 30 RPM for 15 minutes for coagulation and flocculation. Stop stirring; raise the stirring paddles; and allow the settling of the floc for 5, 15, 30, and 60 minutes.

Carefully draw the sample below the surface (to avoid floating solids) and above the sediments for each interval and determine the turbidity. Suppose the test results show that the optimum dose is 6 mg/L after 15 minutes of settling. Start treating the water with 6 mg/L dose and make an adjustment around this dose as needed.

pH

pH, hydronium ion index, is the measurement of acidity (H^+). Acidity in water is usually due to carbon dioxide (CO_2) from rain water, mineral acids, chlorine, and heavy metal salts, such as alum. pH is an important parameter in the water utility. It is used to determine the condition of water for proper coagulation, softening, and stabilization.

Principle and Procedure. The potentiometric or pH meter method is a convenient and accurate method for this test. The meter measures the voltage difference developed between the electrodes due to hydronium ions in the solution. This differential is read as the pH of the sample. Testing procedure varies somewhat with different brands and models of the pH meter. Follow the instructions provided by the manufacturer. Proper calibration by using an appropriate buffer solution (solution with the stable pH), such as 6 or 9, is important. Rinse the electrode with the deionized (DI) water and wipe with a soft paper to remove the DI water. Immerse the electrode in the sample and swirl the sample. Record the pH after the reading is stabilized. The electrode should be rinsed with DI water after every use. Leave the electrode in the DI water when the meter is not in use.

Alkalinity

Alkalinity of water is its capacity to neutralize acidity. Carbonates, bicarbonates, and hydroxides are the most common forms of alkalinity in natural waters. These chemicals are mostly compounds of calcium and magnesium coming from mineral deposits such as limestone and dolomite. Industrial discharges can also cause alkalinity. Bicarbonate alkalinity is present between pH 4.3 and 8.3. Carbonate and bicarbonate alkalinity is present between pH 8.3 and 9.4, and carbonates and hydroxides are present between pH 9.4 and 14. *Alkalinity does not exist below pH 4.3.* Alkalinity test is important to determine proper coagulation and the stability of water.

Principle and Procedure. The commonly used method is titration. Alkalinity of a sample is determined by titrating the sample with 0.02 normal (N) sulfuric acid (H_2SO_4). Results are expressed in mg/L as calcium carbonate. This normality simplifies the calculations of results by giving 1 mg/L of alkalinity for 1 mL of titrant per liter of sample.

Alkalinity is expressed as P (phenolphthalein an indicator) when hydroxides and carbonates are present; they produce a pink color with the indicator phenolphthalein. Alkalinity is called M (methyl orange) when carbonates and bicarbonates are present. One half of carbonate alkalinity is titrated as P and other half as M. The sum of P and M is known as T, or total alkalinity. For an alkalinity test, water sample should be free of turbidity and color because they obscure the indicator color. The chlorine residual above 1.8 mg/L interferes by bleaching the color of the indicator.

Carefully measure the appropriate volume of the sample (normally 50 mL), and neutralize the residual chlorine, if required, by adding two drops of 10 percent sodium thiosulfate. Add two drops of phenolphthalein indicator and swirl. If the sample turns pink, hydroxide, carbonate, or both of them are present. Titrate with sulfuric acid until the pink color disappears. Mark the buret reading (mL used) as "P." If there is no color with phenolphthalein, then P alkalinity is 0, which means hydroxides and carbonates are absent. After P alkalinity titration, add 3 to 6 drops of the indicator methyl orange or bromcreosol green methyl red to the same sample. If M alkalinity is present, the sample will turn yellow with methyl orange and purple with bromcreosol green methyl red. These changes in color indicate the presence of carbonates, bicarbonates, or both. Titrate with sulfuric acid to orange when methyl orange is the indicator and to salmon color for the indicator brom-creosol green methyl red. Record the reading from the start of the test to the end as T. The difference between T and P is the M alkalinity. Calculate the alkalinity as mg/L of calcium carbonate by multiplying the total number of mL of the titrant (T) with factor 20, because 50 mL sample is 1/20 of 1L, or 1,000 mL. Similarly, for 25 mL sample the factor is 40, and for 100 mL it is 10.

Table 3-1 gives the types and calculations of alkalinity for different titration results.

Example. Suppose four samples are tested for alkalinity. The first sample has no P alkalinity and total alkalinity is 225 mg/L; the second sample has 50 mg/L P and 50 mg/L M alkalinity; the third sample has a total alkalinity of 130 mg/L with P alkalinity of only 10 mg/L; and the fourth sample has all P alkalinity, and total is 75 mg/L. Determine the type of alkalinity in each sample.

1. Sample alkalinity = 225 mg/L of bicarbonate alkalinity because there is no P alkalinity.
2. Sample alkalinity = 100 mg/L of carbonate alkalinity, because P is 1/2 of T.
3. Sample has P alkalinity less than half of the total alkalinity; therefore, it has carbonate and bicarbonate alkalinity.
   - Carbonate alkalinity = 2P or 2 × 10 mg/L = 20 mg/L
   - Bicarbonate alkalinity = T − 2P = 130 mg/L − 20 mg/L = 110 mg/L
4. Sample alkalinity = 75 mg/L of hydroxide as there is no M alkalinity.

Hardness

Hardness of water is the total concentration of calcium and magnesium ions expressed as calcium carbonate. Hardness is caused by the soluble bicarbonates, sulfates, nitrates, and chlorides of calcium and magnesium. Hardness,

Table 3-1   Types and Calculations of Alkalinity

Titration Results	Hydroxide	Carbonate	Bicarbonate
P = 0	0	0	T
P = less than 1/2 T	0	2P	T − 2P
P = 1/2 T	0	T	0
P = more than 1/2 T	2P − T	2T − 2P	0
P = T	T	0	0

due to bicarbonates, is known as *carbonate hardness.* When hardness is due to other compounds, it is known as *noncarbonate* hardness. The carbonate and noncarbonate hardness are also known as temporary and permanent hardness, respectively. Other bivalent and trivalent (Fe^{+2} and Al^{+3}) metal ions, such as iron, aluminum, manganese, and zinc will also cause hardness, but their concentrations are mostly insignificant.

Principle and Procedure. The most commonly used method is titration. Calcium and magnesium ions are titrated with disodium ethylenediamine tetra acetate (EDTA). Hardness-causing ions form a chelated compound with EDTA, and the color of the indicator eriochrome black T is changed from wine red to blue at the end point.

Measure 50 mL of sample, buffer it, add indicator, and swirl to mix. If the sample turns wine red, then hardness is present. Titrate the sample with 0.02 N EDTA until the color changes to blue. To avoid overtitrating, pour EDTA drop by drop as soon as the color starts changing to purple. After the purple color, it takes only a few more drops of the titrant to reach the end point. After titration the color may change back to purple due to the reversal of the reaction. For this reason, the first reading is considered a rough reading. The second titration is done carefully to have an accurate reading. For a 50 mL sample, multiply the number of milliters of titrant used with 20 to determine mg/L of hardness as calcium carbonate.

Principle and Procedure for Calcium Hardness. A calcium test is required to determine the stability of the water. Calcium is present in the natural waters at levels ranging from 0 to several hundreds mg/L. Another source of calcium in water is the use of lime (CaO or Ca $(OH)_{2)}$ for softening purposes.

This method is a modification of the total hardness EDTA titration method. The pH of the sample is raised to 12–14 to prevent the magnesium interference. Eriochrome blue-black R is used as an indicator that turns red if calcium is present, and changes to sharp blue at the end point. The alkalinity of the sample above 300 mg/L obscures the end point. This problem is resolved by diluting the sample.

Relationship between Alkalinity, Carbonate and Noncarbonate Hardness. Normally, the alkalinity in water is due to carbonates and bicarbonates of calcium and magnesium. When testing two different parts of the same compounds in the hardness and alkalinity tests, both results are expressed as calcium carbonate. Therefore, the total alkalinity of a sample is equal to its carbonate hardness. Thus noncarbonate hardness is the difference between total hardness and alkalinity.

$$\text{Calcium carbonate} = \underset{\text{Hardness}}{Ca^{+2}} \qquad \underset{\text{Alkalinity}}{CO_3^{-2}}$$

Example: Suppose the alkalinity of a water sample is 100 mg/L, and its total hardness is 125 mg/L:

$$\text{Carbonate hardness} = \text{Total alkalinity} = 100 \text{ mg/L as } CaCO_3$$

$$\text{Noncarbonate hardness} = 150 \text{ mg/L} - 100 \text{ mg/L} = 50 \text{ mg/L as } CaCO_3$$

Disinfection Residual

Chlorine is one of the most effective disinfectants and is quite commonly used for water disinfection. Chlorine, combined with ammonia, forms chloramines, which are called *combined residual chlorine.* Total residual chlorine is the sum of the free residual chlorine and combined residual chlorine.

Principle. The most commonly used method is an amperometric method, which is a modification of the polarographic (automatic measuring and recording) principle. It uses 0.00564 N phenylarsine oxide (PAO) as a titrant. By using this normality, 1 mL of the titrant equals 1 mg/L of chlorine for the sample volume 200 mL. PAO is quite stable and available from venders in a form of accurate strength and ready for use. A special amperometric cell is used to detect the end point. Free residual chlorine is titrated at pH 7 and combined residual at pH 4.

Procedure. Rinse the titration cell of the analyzer by filling the cup with a sample and turning on the mixer. Discard the sample and refill the cup to the 200 mL mark with a grab sample, and turn the mixer on. Add 1 mL of phosphate buffer to adjust the pH to 7 and adjust the needle to the middle of the scale. Add small amounts of PAO until deflection of the needle to the left stops. This is the end point for free residual chlorine; the milliliters of the titrant used are equal to free residual chlorine as mg/L. Add to the same sample the potassium iodide (KI) solution and 1 mL of the acetate buffer to adjust the pH to 4. If the combined residual chlorine is present, the needle will move to the right. Adjust the needle to the middle of the scale and titrate until the deflection to the left stops. This second reading is the combined residual chlorine. Assume the first end point comes after 2 mL and the second comes after 3 mL. Then 2 mg/L is free; 1 mg/L is combined; and 3 mg/L is total residual chlorine. After testing, discard the sample and rinse the cell with tap water and turn off the mixer.

Colormetric method by using diethyl phenylenediamine (DPD) is a common field method. The colormetric method is based on the principle that the darker the color of the indicator, the higher is the concentration, and vice versa. Being less accurate, it is not used for compliance purposes. It is quite simple and an easy test. Free-residual chlorine reacts with DPD to produce a pinkish red color. The color intensity is matched with the standard color on the color wheel to determine the amount of the residual chlorine. For total-residual chlorine, use potassium iodide along with DPD to liberate iodine from the combined residual chlorine. Iodine reacts with DPD to produce the red color. Use of a color comparator test kit by the Hach Company is quite common. The company provides powder pillows of various chemicals used in the test and simple instructions for testing.

Coliform Bacteria Tests

Bacteriological quality of water is important to determine the degree of disinfection and possible presence of waterborne pathogens. Bacteria, being small, are present almost everywhere, such as in air, water, and on lab equipment. Therefore, all equipment and handling is done in a sterile environment to ensure the accuracy of data.

Preparation. All glassware to be used in the bacteriological tests should be thoroughly cleaned in a glass-washing machine and sterilized. If a sample has chlorine, then use two drops of 10 percent sodium thiosulfate solution per sample bottle before sterilizing. For thorough sterilization in an autoclave, bottles should be loosely capped to allow the steam to penetrate. Caps are covered with aluminum foil to prevent the contamination by hands while handling the sample. Sterilize the capped bottles in an autoclave for 15 minutes at 15 psi pressure, which equals 121°C. Cool the sterilized bottles to room temperature and tighten the caps.

If caps are tightened when bottles are still hot, condensation of the steam inside the bottle will create a vacuum, which will make the cap hard to open. Moreover, when opened, the bottle will suck air and will be contaminated before collecting a sample.

Growth Media. Media are the food for the bacteria to culture them in the laboratory. Different bacteria have different food requirements; therefore, each medium will allow certain types of bacteria to grow. Media are either liquid, known as *broths,* or semisolid (gelatinous), which are called *agars.* Agar, an extract of seaweeds, is used as a solidifying agent in the media to facilitate the growth of individual bacteria into colonies by keeping the media solid at incubation temperature below 45°C. The number of colonies corresponds to the number of individual bacteria in the sample.

A medium is called *selective* if it contains certain chemicals that allow only certain types of bacteria to grow. For example, brilliant green bile broth (BGB) allows only fecal coliforms to grow from the coliform group. A medium is known as *differential* if different types of bacterial colonies are differentiated by different colors, shapes, and sizes. For example, M endo broth gives coliform colonies the golden green metallic sheen.

Preferably, media should be freshly prepared daily. Prepare and sterilize media according to the instructions from the supplier. Dilution water, for preparation of media, is buffered to pH 7.2 with phosphate buffer solution.

Label all the sample bottles and petri dishes properly. Petri dishes should be labeled on the underside, as they are incubated upside down.

Sampling. Proper sampling is important. Avoid contamination of a sample from hands, air, and sneezing. To collect a representative grab sample from a tap, open the faucet fully and allow the water to flow for 2 to 3 minutes to flush out the stagnant water. If the sample is collected from a river, stream, or lake, hold the bottle upside down below the surface to avoid any floating material from entering the bottle, turn the bottle upward facing the current and fill it. (Hold the cap upside down in the other hand.) If there is no current, push the bottle forward to fill it. If the sample comes from a well, run the well water for 5 minutes before collecting the sample.

Collect at least 100 mL of sample by filling only two thirds or three quarters of the sample bottle or container to leave enough space for the proper mixing of the sample before testing. Take the sample immediately to the laboratory or ship it to its testing destination. During the holding time, which should be not more than eight hours for coliform bacteria, keep the samples at 5 to 10°C to avoid any change in the bacterial density.

Coliform Test. To ensure the absence of waterborne pathogens, the water is tested for coliform bacteria. Coliform bacteria are present in human wastes and in soil contaminated with human wastes. These bacteria in human wastes are known as *fecal coliform* bacteria. Those in the soil are called *nonfecal coliforms*. Fecal coliforms are represented by *Eschrechia coli* (*E. coli*), and nonfecal coliforms are represented by *Enterobacter aerogenes* (*E. aerogenes*). Both fecal coliforms and nonfecal coliforms are called the *total coliform group*. This group is used as an indicator of the presence of human wastes in water and the possible presence of waterborne pathogens.

For compliance, use a state-approved sampling site plan of the distribution system. The number of samples/month is based on the population being served. Testing methodology for coliform bacteria has four different techniques:

1. Membrane filter technique
2. Multiple tube fermentation (MTF)/most probable number (MPN) technique
3. Presence–absence technique
4. Defined substrate technology technique (e.g., Colilert)

Membrane filter technique is the filtration of the sample through a 0.45 micrometer pore-size filter, which retains the coliform bacteria that are then provided with a selective and differential media pad in a tight-fitting petri dish (see Figure 3-2). M. endo broth is used for total coliforms, and M FC broth is used for the fecal coliforms. A pad is soaked with about 2 mL of media, and the filter is placed on it. The dishes are incubated upside down (to prevent drying of the filter) at 35.5 ± 0.5°C for total coliforms and 44.5

Figure 3-2   Placing a Membrane Filter on a Pad Soaked with Medium in a Petri Dish (*Source: M-12—Simplified Procedures for Water Examination.* American Water Works Association, 1964.)

± 0.2°C for fecal coliforms for 24 ± 2 hours. Each coliform bacterium multiplies every half hour and forms a colony. Total coliform bacterial colonies have a distinct golden green metallic sheen, and fecal coliform colonies are blue. This technique gives both qualitative and quantitative results. Therefore, it is a commonly used technique in most water treatment laboratories. Results are reported as colonies per 100 mL of sample.

For accuracy, use the results of the filters with a colony count of 20 to 80 for coliforms and 20 to 60 for fecal coliforms (fecal coliform colonies are bigger). If all filters have a colony count less than 20 each, then total all such counts and compute the results from total volume of the sample. If all filters have a colony count higher than the upper limit (80 for coliforms and 60 for fecal coliforms) or more than 200 total colonies (all kinds), then report as too numerous to count (TNTC). If colonies are fused and not well defined, then report as confluent and repeat the test.

The *multiple tube fermentation technique* is based on the principle of selecting coliform bacteria by using a selective medium and culturing them to produce gas. It is a qualitative test.

A set of five or multiple of five culture tubes (test tubes with an inverted small tube to trap the gas) containing the required selective media is inoculated with the water sample. Sample and media are mixed for the proper dispersal of microbes. These tubes are then incubated at the required temperature for 24 to 48 hours. Coliform bacteria produce gas by fermenting lactose. Presence of gas in the insert tube indicates a positive test (see Figure 3-3).

This test is divided into three parts: presumptive, confirmed, and completed tests. Laurel tryptose broth, brilliant green bile broth (BGB), and lactose broth are the respective media for these parts. The presence of gas in the laurel tryptose broth shows a positive presumptive test indicating the *possible pres-*

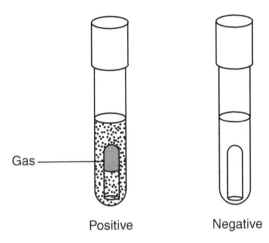

Gas

Positive   Negative

Figure 3-3   Culture Tubes

ence of coliform bacteria; the presence of gas in the BGB tubes means a positive confirmed test that *confirms* the presence of coliform bacteria; and a completed test *reconfirms* their presence by showing the presence of gas, gram negative (red color), and nonsporulating bacilli. The last three characteristics of coliform bacteria are determined by preparing a slide of the culture, staining it with gram stain, and observing the culture under a microscope. It takes 2 to 5 days to have the final results.

Use EC broth and incubation at 44°C for the fecal coliforms completed test. The density per 100 mL sample is estimated from the MPN table, which is based on the number of the positive tubes. This technique, being lengthy and laborious, is less popular than the membrane filter technique. It is normally used as a back-up test with the membrane filter technique.

Presence–absence technique is a qualitative test based on the incubation of 100 mL of water sample and laurel tryptose broth to indicate the presence of coliforms by fermenting the media. The presence of yellow color after 24 to 48 hrs incubation at 35 ± 0.5°C indicates a positive presumptive test for coliform bacteria, which is confirmed by using BGB broth. It is a modification of the MTF technique to encourage even a single coliform bacterium to grow and give a positive result.

Defined substrate technique is based on the principle that each organism has some specific enzyme. An *enzyme* is a chemical that accelerates the rate of a specific reaction of a chemical called *substrate,* without undergoing any change in itself. The presence of the organism is tested by using the specific substrate for that particular enzyme. If a reaction takes place, the organism is present and the test is positive, and vice versa. An indicator chemical is used to show the reaction. This test is a biochemical test for specific organisms. Enzyme beta-galactosidase is specific to the total coliforms, and enzyme beta-glucuronidase is found only in *E. coli.* In the Colilert test the medium contains substrate bonded with indicators ortho nitrophenyl-beta-d-galactopyranoside (ONPG) for total coliforms and 4-methylumbelliferyl-beta-d-glucuronide (MUG) for *E. coli*. This medium is known as ONPG-MUG, or minimal media, MMO-MUG. When the sample is incubated with these chemicals in the media, coliform bacteria produce a yellow color that fluoresces in ultraviolet light if *E. coli* is present.

Procedure. Add 100 mL of sample to the container with Colilert media and incubate at 35 ± 0.5°C for 24 hours. Check for yellow color and fluorescence with 366 nm wavelength UV light. The density of the coliform bacteria can be determined by using the multiple tube technique with ONPG-MUG and applying the MPN table. This test gives total coliform and *E. coli* results within 24 hours, which is a useful tool for the water utility under crisis conditions (like possible cross connection) for a quick response. This fast, sensitive, and specific technique is becoming popular with water utilities. The only drawback is that it is still an expensive technique.

For more information, refer to coliform bacteria in Chapter 17.

Heterotrophic or Standard Plate Count (HPC)

This test gives the total count of almost all types of bacteria in the water sample that can grow on a general medium called the *standard plate count agar* or *nutrient agar*. A count less than 500 colonies/mL of the sample means that the water is properly disinfected, and vice versa. Furthermore, a count of higher than 500 colonies/mL interferes with the growth of total coliform bacteria. It is a supplementary test for process control. This test uses only 0.1 to 1 mL of a sample. Use a straight sample for treated water and a dilution for raw waters.

Principle. Standard plate count agar allows most of the bacteria (heterotrophic baceria) in the water to grow and form colonies of different shapes and sizes, in 48 hrs. at 35°C, incubation temperature. Colony count/mL of sample determines the quality of water. The higher the count, the more polluted the water, and vice versa. This test is used to determine the quality of source water, level of disinfection in the proces control, and adequate disinfection of water in the distribution system. If the number is more than 500/mL of sample from the distribution system, the system is deficient of disinfectant.

Procedure. Mix a sample or dilution thoroughly in a capped bottle and transfer 0.1 and 1 mL to two sterilized 100 mm × 15 mm petri dishes, respectively. Add to each dish 10 to 15 mL of the tryptone glucose extract agar (standard plate count agar) that has been liquefied, sterilized, and cooled to 44 ± 1°C, and rotate the dish gently to mix the sample and medium evenly. Allow the mixture to solidify; invert the plates, and incubate them at 35 ± 0.5°C for

48 ± 3 hours. Use a Quebec Colony Counter and count all bacterial colonies (even pin/point size). Report results as number of bacteria per milliliter of sample.

Valid and accurate data from the quality control lab ensure the adequate water quality and compliance with the SDWA.

QUESTIONS

1. What is the importance of a jar test? State three important water treatment requirements where jar testing will be useful.

2. Define the terms *NTU* and *threshold odor numbers.*

3. What causes turbidity? Why is turbidity one of the most important parameters to determine the potability of the drinking water?

4. **a.** Define *titrant.*
   b. Mostly applied normality of a titrant is that which has 1 mL of the titrant equivalent 1 mg of the titrated substance. T or F
   c. What normalities of H_2SO_4 and EDTA titrants are used for the alkalinity and hardness tests?

5. Explain the importance of alkalinity and hardness tests. Why is the total alkalinity equal to carbonate hardness?

6. Suppose a water sample has total alkalinity 50 mg/L and total hardness 125 mg/L, both as $CaCO_3$. Calculate the carbonate and noncarbonate hardness of this sample.

7. Calculate OH^-, CO_3^{-2}, and HCO_3^- alkalinities of the following samples:
   a. P alkalinity is 50 mg/L and total alkalinity is also 50 mg/L.
   b. P alkalinity is 50 mg/L and total alkalinity is 100 mg/L.
   c. P alkalinity is 0 and total alkalinity is 75 mg/L.

8. **a.** Alkalinity does not exist below pH 4.3. T or F
   b. Alkalinity can be present in a sample below pH 7. T or F

9. **a.** Chlorine produces dark red color with indicator DPD. T or F
   b. Amperometric method for chlorine residual uses PAO as a titrant with _____ N.
   c. Milliliters of PAO used to reach the end point are equal to the mg/L of the residual chlorine. T or F

10. Define MTF, MPN, and HPC.

11. State the basic principle applied in the membrane filter, multiple tube fermentation, presence and absence, and defined substrate techniques for coliform testing.

12. **a.** Heterotrophic plate count is the same as the standard plate count. T or F

    b. What is the significance of the HPC test?

4

PRETREATMENT

Pretreatment is the preparation of raw water for an adequate treatment before it reaches the treatment plant. Especially for surface waters, this process involves removing undesirable objects and contaminants:

- It screens any large floating objects such as sticks, leaves, logs, rags, fish, and zebra mussel.
- It removes most of the turbidity formed of sand, grit, organic matter like humus and fecal matter, and colloidal particles.
- It controls tastes, odors, and synthetic organics.
- It controls THM precursors.
- It kills waterborne pathogens and other bacteria.

The purpose of this phase is to reduce the treatment load on the main plant as much as possible and to reduce wear and tear on the equipment.

Before treating water, it is very important to have a thorough knowledge of its source, quality, and available quantity, including seasonal variations. In addition, it is important to have a thorough understanding of the treatment facilities. The quality of water varies from source to source, from region to region, and from season to season. Therefore, each treatment plant is unique in its design and operation. Generally, a treatment plant is a multibarrier system that provides a barrier for waterborne pathogens and other contaminants at each step of the treatment.

A typical municipal water treatment plant with sedimentation, filteration, and disinfection is known as a *conventional treatment plant*. It generally includes some or all of the following parts:

- Intake
- Bar screen
- Rapid mix
- Presedimentation basins
- Transmission line
- Rapid mix
- Coagulation and flocculation basins
- Primary sedimentation basins
- Rapid mix
- Final sedimentation basins
- Filters
- Clear well
- Reservoir
- High service pumps

These parts together are called the *treatment train.* Figure 4-1 diagrams a typical treatment.

PRETREATMENT PROCESSES

Screening

Screening serves to remove large floating objects such as sticks, leaves, logs, rags, fish, and zebra mussel, algae, and insect larvae. There are two main types of screens: bar screens and microstrainers.

- *Bar screens.* These are fixed or movable screens with bars, commonly, 1 to 3 inches apart. They are the first treatment process to remove large objects. The screenings are removed manually or mechanically and disposed off.
- *Microstrainers.* They are finer screens used after bar screens to remove finer particulate matter, such as algae. They are used by themselves when the source water (good quality reservoir) does not have large objects.

Rapid Mixing

Rapid mixing chambers are used for mixing chemicals before further pretreatment. After bar screening, the low lift pumps deliver the water from the source to the rapid mix, which has the provision of feeding chemicals such as polymer, potassium permanganate, and activated carbon. Chemicals are rapidly incorporated into the water in the rapid mix in less than 45 seconds. After rapid mixing, water flows to the presedimentation basin. These basins are large basins normally 15 to 20 feet deep for gravity settling of turbidity.

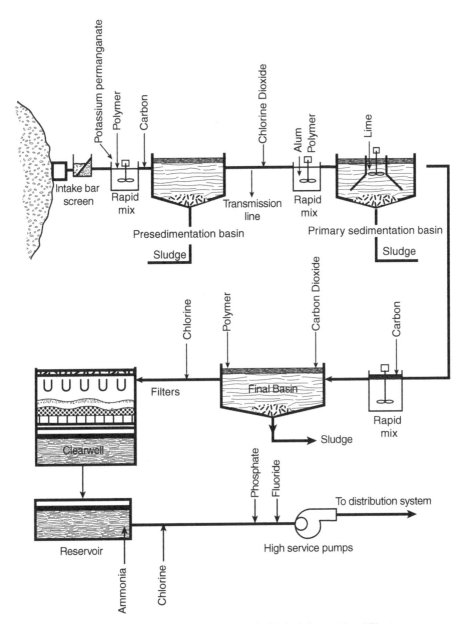

Figure 4-1 A Diagrammatic Sketch of a Typical Conventional Plant

Presedimentation

Presedimentation is the settling out of the turbidity particles, such as sand, grit, and organic matter such as humus and fecal matter. It is mostly a plain sedimentation (without using any coagulant like alum); however, a small dose of a polymer is often applied for an effective turbidity removal. Turbidity particles settle to the bottom of the presedimentation basin after an adequate detention time, which is mostly one to two hours. Settled particles are known as *sludge,* which is removed periodically by an automatic sludge-removal system. The removal time and interval can be set as required. An adequate sludge withdrawal from the presedimentation basin is important. Remove the sludge more for higher turbidities and less for lower. If sediments are not removed adequately, the turbidity particles will pass out with the flow. Turbidity is reduced by about 85 percent or better, which makes the rest of the treatment easier.

It is a good idea to run a jar test for different turbidity levels in the source water. For example, a jar test showed that the following polymer doses were effective for different levels of turbidity in river water, as shown in Table 4-1.

The optimum dose for 700 nephalometric turbidity units (NTU) turbidity and higher was 7 mg/L. Each source of water needs guidelines, since differences are usual.

Taste and Odor Control

For this purpose, powdered activated carbon and potassium permanganate are used in the rapid mix. Removal takes place in the presedimentation basin. Activated carbon adsorbs (acquires on the surface) taste- and odor-causing substances and synthetic organics, such as atrazine. Potassium permanganate in small amounts (0.5 to 1 mg/L) removes tastes and odors by oxidizing them. Furthermore, it reduces the waterborne pathogens and coliform densities; it also controls zebra mussel. Potassium permanganate is popularly termed *purple magic.*

Predisinfection

Disinfection is the inactivation of waterborne pathogens. Predisinfection is the first disinfection process. Settled water, flow from the presedimentation

Table 4-1 Effective Polymer Doses for Turbidity

Raw Water Turbidity	Polymer Dose
30–50 NTU	0.5 mg/L
50–100 NTU	1 mg/L
Above 100 NTU	1 mg/L for every 100 NTU

Table 4-2 Pretreatment Problems and Their Solutions

Problems	Possible Causes	Possible Solutions
Presedimentation basin turbidity increases when source water turbidity and other conditions are unchanged.	Improper sludge removal.	Check sludge blanket level. If too high, increase sludge removal time and frequency.
	Sludge scrapers are stopped.	Check scraper movement to be sure it is working.
	Polymer feed pump does not feed adequately.	Check feed pump, setting, and polymer strength. To eliminate polymer diluting problem, feed straight polymer. Adjust feeder for required dose.
	Inadequate polymer dose. Both under and overfeeding can cause high turbidity. Lower dose does not coagulate all turbidity particles. Excessive overfeeding increases water viscosity, which interferes with settling.	Run jar test; determine proper dose of polymer; apply it.
Presedimentation basin has high turbidity after heavy winter rain.	Too much sand and silt in water. Turbidity comes from sand and silt caused by ice chunks in the river or stream. Rolling movements of river ice stir bottom sediments that will cause high turbidity.	Remove sludge more frequently for longer periods to get rid of heavy solids. This also increases detention time in basin. Check sludge density. Withdrawal period should stop when sludge is thin and watery to reduce unnecessary water wastage.
	Insufficient detention time.	Allow longer detention time to accommodate for almost-freezing water, which is heavy and subject to slow sedimentation.
	Inadequate polymer dose.	Run jar test to determine optimum polymer dose; too much or too little polymer may be the cause. Avoid feeding polymer over optimum dose.

Table 4-2 *(Continued)*

Problems	Possible Causes	Possible Solutions
Presedimentation basin has high turbidity after heavy summer rain.	Inadequate sludge removal. Sand and silt settle rapidly. Removal of an adequate amount of sludge is important.	Increase frequency and withdrawal time of sludge to accommodate for high temperatures which increase rate of sedimentation.
	Inadequate polymer treatment.	Run jar test to determine optimum dose. Apply optimum dose; fine tune dose to achieve best results. A lower polymer dose in summer is better.
In summer, gas bubbles rise to surface of source water and leave a small patch of oil-like film.	Anaerobic decomposition of sludge at bottom of water body. Under drought conditions, the oxygen supply at bottom of water body is depleted.	To be sure it is not an oil spill, inspect source water; run odor test, and check water for unusual contaminants.
Higher disinfectant demand of presedimentation basin effluent, but turbidity removal is satisfactory.	High bacterial densities.	Check coliform count on the settled water. High bacterial count will cause a high demand. Increase predisinfectant dose as required.
	Discharge of raw sewage from farm land or other town upstream.	Check raw water coliform density. If up, there will be higher demand for disinfectant. Increase disinfectant dose. Check with municipalities upstream.
	Inadequate potassium permanganate ($KMnO_4$) dose.	If $KMnO_4$ is used in presedimentation basin, its dose may be too low. Adjust $KMnO_4$ dose 0.5–1.0 mg/L. $KMnO_4$ will reduce demand by killing microorganisms by providing partial disinfection. Too much $KMnO_4$ can cause pink water condition if detention time is short.

Observation	Possible Cause	Remedy
Higher coliform and heterotrophic plate count densities are in the presettled water.	Inadequate predisinfection.	Check disinfection dose and adjust as required. Also, check for leaks in disinfectant line. There might also be an underground leak.
	Deficient $KMnO_4$ dose.	Check $KMnO_4$ feed system and apply required dose. Control bacterial densities before they cause problems at treatment plant.
Fishy and earthy-musty odors from presedimentation basin, but there is no problem with source water.	Sludge decomposing at the bottom of the basin.	If scrapers don't scrape sludge deep enough, decomposition will occur. Clean basins and wash with high pressure hose at least twice/year. Adjust scrapers. Practice good housekeeping.
	Any overflow from a sludge lagoon.	Check for any such problem and correct. Any stagnant water with sediments in bottom should be noted.
High atrazine in the presettled water.	High atrazine in the source water. This pesticide is applied to control weeds in wheat and sorghum fields. It enters source water after heavy rains in spring and June and July.	Monitor atrazine level regularly in source water. Run jar test. Determine optimum dose of carbon and adjust accordingly.
	Low carbon slurry concentration.	Check slurry concentration of carbon. Adjust dose corresponding to slurry concentration.
	Incorrect carbon feed setting.	Monitor feed rate and check setting for slurry concentration.
	Mixer of carbon slurry tank is not working.	Mixer should always be on to keep the carbon suspended.

basin, is disinfected by an appropriate disinfectant. If the water has a low potential of THM formation, then a free-residual chlorine dose generally less than 1 mg/L is used. If the potential is high, then chloramine, chlorine dioxide, or ozone is applied. For chloramine applications, ammonia is used before the chlorine feed to minimize THM formation. Ammonia reacts with chlorine before the THM precursors. If there is a long transmission line from the pretreatment facility to the treatment plant, chlorine or chloramines may satisfy most of the disinfection requirements. If the transmission line is short, then the use of chlorine dioxide or ozone would be the best choice.

Normally, pretreated water has a very low turbidity and is almost completely disinfected. Requirement and degree of pretreatment depend on the quality of source water. Water from high-quality reservoirs in the mountains may need only microstraining. Pristine streamwater may need only barscreening followed by microstraining. Pretreated water is transmitted to the main treatment plant for further treatment.

Whether the source of water is a natural lake, a manmade reservoir, a pristine stream, or a muddy river, pretreatment is the first important step to make the water treatment easier, effective, and economical.

Sedimentation, disinfection, and taste and odor control are discussed in detail in their respective chapters (Chapters 6, 10 and 11).

Pretreatment Problems and Their Possible Solutions

For pretreatment problems and possible solutions, refer to Table 4-2. This table is a useful reference tool for identifying and correcting pretreatment problems.

QUESTIONS

1. Write three main uses of a public water supply.

2. What is the main purpose of pretreatment?

3. Why is ammonia fed before the chlorine for chloramination?

4. What is the other use of $KMnO_4$ besides taste and odor control?

5

COAGULATION
AND FLOCCULATION

Coagulation and flocculation convert nonsettlable turbidity particles into sett-lable form for their effective removal by gravity.

After presedimentation, these particles are mostly colloidal type. Colloidal turbidity particles are too small (1–100 nm) to settle by gravity. They stay suspended and cause turbidity. Mostly, they are negatively charged. Their removal is accomplished by using substances that make them clump together to form large and heavy particles known as *floc* that will settle. These substances are known as coagulants. A *coagulant* is an electrolyte that provides *cations* (positively charged ions) to precipitate out the negatively charged colloidal turbidity particles. As a rule, the higher the charge on the cation, the more effective is the coagulant. Therefore, commonly used coagulants are aluminum and ferric compounds that provide Al^{+3} and Fe^{+3} cations, respectively. Some other substances are often used to facilitate the coagulants; those are known as coagulant aids. This treatment phase is the second barrier to remove turbidity, waterborne pathogens, and other contaminants. It consists of three parts: rapid mixing, coagulation, and flocculation.

RAPID MIXING

Rapid mixing is the fast and thorough mixing—*flash mixing*—of the various chemicals, such as coagulants and coagulant aids, with water for their proper chemical reactions. It has only 30 ± 15 seconds detention time in a small tank. Rapid mixing disperses the chemicals immediately to reach their targets and start their precipitation. Precipitation (separation of a solid from a liquid) in water treatment is known as *coagulation*. It is the start of the removal of colloidal particles. Rapid mixing is followed by coagulation.

COAGULATION

Coagulation is the precipitation of the colloidal turbidity particles, coagulants, and coagulant aids.

Steps of Coagulation

Coagulation occurs in three steps. First, Al^{+3} or Fe^{+3} ions attract a considerable number of negative colloidal turbidity particles. Second, due to aggregation, they form small clumps, called *micro-floc*. Third, micro-floc, due to its positive charge, still attracts negative ions such as alkalinity (OH^- from lime) and floc compounds precipitate due to their low solubility. See Figure 5-1.

Chemical reactions:

$$Al_2(SO_4)_3 + 3Ca(OH)_2 \rightarrow 2Al(OH)_3\downarrow + 3CaSO_4$$

Alum   Lime   Floc

$$Fe_2(SO_4)_3) + 3Ca(OH)_2 \rightarrow 2Fe(OH)_3\downarrow + 3CaSO_4$$

Ferric sulfate   Floc

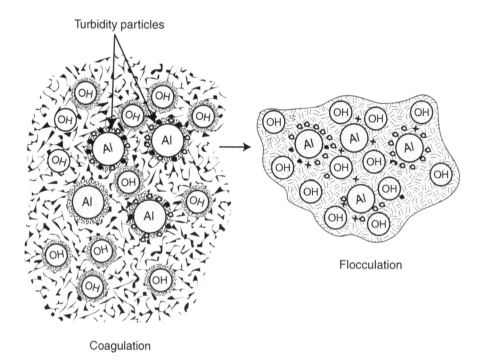

Figure 5-1   Coagulation and Flocculation Process

Factors Affecting Coagulation

1. *Coagulant.* Different sources of water need different coagulants, such as the following:
   - *Filter alum,* $Al_2 (SO_4)_3 \cdot 14 H_2O$ (aluminum sulfate), a powder, is one of the most commonly used coagulants. It is a good coagulant for hard waters with high alkalinity and pH 5.5–8.0. It is also available as liquid alum (Al_2O_3). Both provide Al^{+3} ions in the water. Normally, coagulation needs 1 mg/L of alum for every 5 NTU turbidity up to 30 NTU; and above that, 1 mg/L for every 10 NTU.
   - *Activated alum* is alum with about 9 percent sodium silicate. It works as coagulant and coagulant aid.
   - *Black alum* is alum containing activated carbon. It is applied for certain water with carbon adsorption requirements.
   - *Ferric sulfate* $(Fe_2(SO_4)_3)$ is the second most commonly used coagulant, which works for water with a pH range from 5 to 11.
   - *Ferrous sulfate* $(FeSO_4)$ is useful for water with high pH (8.5–11).
   - *Chlorinated copperas,* a mixture of ferric sulfate and ferric chloride $(FeCl_3)$, has also been used for water with a pH range from 5 to 11.
   - *Sodium aluminate* $(NaAlO_2)$ is useful for hard water as it works both as a softener and a coagulant. Its solution is alkaline with pH 12.
2. *Coagulant aids.* These help the coagulation by creating better coagulation conditions, such as proper pH, alkalinity, and particulate nuclei. Some of them act as secondary coagulants.
   - *pH adjusting coagulant aids* include lime, sodium carbonate, sodium bicarbonate, sodium hydroxide, hydrochloric acid, and sulfuric acid. The first four chemicals raise pH and alkalinity and the last two lower it.
   - *Non-pH affecting coagulant aids* are substances that provide particulate matter as nuclei for coagulation (e.g., clay, sodium silicate, and activated silica). They are also called *weighting substances*. These aids are useful when the turbidity is low and is hard to remove.
   - *Coagulating aids acting as secondary coagulants* are polymers, mostly cationic. They attract negatively charged turbidity particles. Depending on the quality of water, anionic or nonionic polymers may work better for certain water.
3. *pH.* Effectiveness of a coagulant is generally pH dependent. Different water requires different coagulants based on its pH. Water with a color will coagulate better at low pH (4.4–6) with alum.
4. *Alkalinity.* It is needed to provide anions, such as (OH^-) for forming insoluble compounds to precipitate them out. It could be naturally present in the water or needed to be added as hydroxides, carbonates, or

bicarbonates, as coagulant aids. Generally, 1 part alum uses 0.5 parts alkalinity for proper coagulation.

5. *Temperature.* The higher the temperature, the faster the reaction, and the more effective is the coagulation. Winter temperature will slow down the reaction rate, which can be helped by an extended detention time. Mostly, it is naturally provided due to lower water demand in winter.

6. *Time.* Proper mixing and detention times are important.

7. *Velocity.* The higher velocity causes the shearing or breaking of floc particles, and lower velocity will let them settle in the flocculation basins. Velocity around 1 ft/sec in the flocculation basins should be maintained.

8. *Zeta potential.* It is the charge at the boundary of the colloidal turbidity particle and the surrounding water. The higher the charge, the more is the repulsion between the turbidity particles, less the coagulation, and vice versa. Higher zeta potential requires the higher coagulant dose. An effective coagulation is aimed at reducing zeta potential charge to almost 0.

Selection of a proper coagulant and a coagulant aid for a water supply is important. A jar test for water should be run to determine which coagulant and coagulant aid are economical and most effective.

FLOCCULATION

Flocculation is the clumping of microfloc particles to form large particles called *floc.* It is achieved by the gentle mixing of coagulated water in tanks known as *flocculation basins* to allow further clumping of the coagulated matter and turbidity particles to form large floc particles. Flocculation basins have slow mixing mechanical paddles, known as *flocculators,* and baffles to provide adequate mixing and low velocity. These basins have a velocity about 1 ± 0.25 ft/sec and detention time of 15 to 45 minutes. The best floc is pinhead size and visible a few feet below the surface of coagulated water. Floc particles are heavy enough to settle to the bottom of basin by gravity. The flocculated water flows to the primary sedimentation basin for the next phase, the sedimentation.

The hardness removal is also achieved at this phase of treatment by using lime and soda ash. If chlorine dioxide is used for predisinfection, then chlorite removal can also be done here by using ferrous ions. For the proper chemical reactions, these chemicals should be applied in the following sequence: ferrous ions, alum, lime, and then soda ash. These topics will be discussed later in their respective chapters.

Refer to Table 5-1 for common coagulation and flocculation problems.

Table 5-1   Coagulation and Flocculation Problems and Their Solution

Problems	Possible Causes	Possible Solutions
Poor floc formation.	Inadequate coagulant dose.	Run jar test; determine optimum dose and increase coagulant dose as required.
	Improper detention time.	Check required detention time by running jar test. Apply needed detention time, if possible, by adjusting flocculators' speed or changing flow rate.
Flakey-feathery floc.	Excess lime. Lime has a low solubility. Excess lime will precipitate as calcium hydroxide and form light floc.	Run jar test; lower lime dose as required.
	Inadequate coagulant dose. Coagulants form heavy floc.	If excess lime dose is desirable, increase coagulant dose until floc quality is improved.
	Improper mixing.	Check rapid mix and mixer speed; adjust as needed.
Poor flocculation when optimum dose of coagulant is used.	Not enough turbidity for an effective flocculation.	Try some weighting coagulant aid like clay or sodium silicate.
Poor floc formation under winter conditions with low water turbidity.	Improper detention time. Low temperature causes slower coagulation which needs longer detention time.	Determine optimum detention time with jar test; apply it.
Inadequate flocculation of yellowish water.	Color of water is due to decomposition of natural organic substances like leaves.	Provide low pH and high dose of coagulant. Alum lowers pH by forming sulfuric acid in the water.
Inadequate flocculation of summer water with low turbidity.	Drought conditions. A lack of proper dilution factor and high concentration of minerals cause poor flocculation conditions.	Run jar test by using alum and a weighting coagulant aid that will increase floc density and rate of coagulation by providing nuclei.
Floc settles in coagulation basins.	Excessive coagulant dose forms heavy floc.	Run jar test to check coagulant dose and lower it as required.
	Weighting coagulant aid dose is too high.	Run jar test with and without coagulant aid to determine if coagulant aid is needed. Lower or stop feeding coagulant aid as required.
	Velocity in basin is too low.	Check velocity and flocculator's speed. Increase velocity as needed since too low velocity allows sedimentation of floc in basins.

QUESTIONS

1. Define these terms:
   a. Rapid mixing
   b. Coagulation
   c. Flocculation
   d. Micro-floc
   e. Floc

2. Name some important chemical coagulants. Why are trivalent cations important for an effective coagulation?

3. Name three important factors affecting coagulation.

4. What kind of water will require alum as a coagulant?

5. Flocculation is the gentle mixing of coagulated water for clumping. T or F

6

SEDIMENTATION

Sedimentation is the settling by gravity of floc from the flocculated water. It is achieved by retaining the flocculated water under quiescent conditions in the large tanks, called *sedimentation basins* or *clarifiers*. An effective sedimentation can achieve up to 99.9 percent reduction in turbidity.

Sedimentation basins are designed to provide quiescent conditions, such as a uniform low velocity, proper detention time, no short circuits, and no surface turbulence to the flocculated water for an effective sedimentation. Velocity varies from 1 to 3 ft/minute. Detention time (the time it takes to fill the tank) ranges from 1 to 6 hours, depending on the design of the basin. A *short circuit* is the flowing of incoming water through the tank, without proper mixing with the water in the basin. Short circuits are generally caused by the stratification of the water in the basin and are common during summer and winter. There should be a proper mixing and baffling of the influent with the water in the basin. Surface turbulence is due to wind action and movements of the equipment. An adequate wind-breaking height of the basin wall above the water surface and the proper maintenance of equipment are very helpful. Flow into the basin is known as an *influent*. Its discharge is called an *effluent*.

Sediments settle to the bottom of the basin, and the effluent is decanted from the top. These basins have inlet and outlet valves for flow control, and the bottom floor slopes to a hopper (pit) for the sludge collection. Sedimentation basins are classified as primary and final sedimentation basins, according to their functions. A *primary sedimentation basin* receives the flocculated water and discharges it to the *final sedimentation basin*.

PRIMARY SEDIMENTATION BASINS

There are two types of primary sedimentation basisn: conventional and high rate.

Conventional Basins

Conventional basins are rectangular or circular. Generally, they are 15 to 20 feet deep to allow proper sedimentation by keeping sludge, light floc (above the sludge), and clear water on the top well separated. Thus, carryover of the floc into the effluent is prevented. Detention time of water in these basins is 4 to 6 hours.

A *rectangular* basin has a horizontal flow from inlet to the outlet (see Figure 6-1). The inlet is at one end and the outlet is at the other. This type of flow is known as a *plug flow*. Generally, length to width ratio is 20 to 1.

Dimensions and proper baffling of these basins allow proper mixing, low velocity, and no short circuits. Sludge is normally collected into a hopper close to the inlet end and is discharged periodically.

A *circular* basin has the radial flow from central inlet. It may have a central chemical mixing zone (see Figure 6-2). Sludge scrapers collect the sludge into a central hopper.

High Rate Basins

High rate basins are designed for a better treatment with high load and less detention time. They are compact units. Detention time is generally 1 to 2 hours, as compared to 4 to 6 hours in the conventional basins. These basins consist of tube settler basins, plate settler basins, and solid contact basins.

Tube Settler Basins. Tube settler basins have tubes installed in them to increase the settling surface and adequate baffling for better sedimentation in a less space.

Plate Settler Basins. Like the tube settler, plate settler basins have plates instead of tubes for a similar function.

Solid Contact Basins. Solid contact basins are compact units with rapid mixing, coagulation, and sedimentation zones in one unit. They are circular basins. A small percentage of previously formed floc is mixed and recirculated

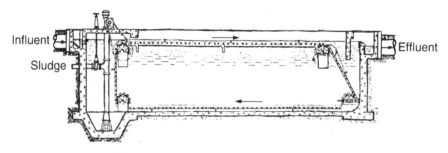

Figure 6-1 A Rectangular Sedimentation Basin

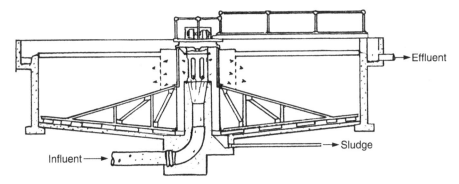

Figure 6-2 A Circular Sedimentation Basin

with the coagulating water in the central part for a fast and economical flocculation. This mixture is called *slurry*. Preformed floc is beneficial by providing the required particulate surface, by providing some of its remaining coagulation capacity, and by acting as a scrubber to remove turbidity from the influent water.

A certain percentage of solids in the mixture are maintained for the optimum performance. Mostly, 3 to 5 percent of 15 minutes settlable solids by volume are the best. To run the settlability test, one should take 100 mL, of a representative slurry sample in a graduated cylinder, mix it gently for 30 seconds, and allow it to settle for 5, 10, and 15 minutes. If the floc settles readily in 5 minutes and the top part of liquid (supernate) is clear, then the treatment is adequate. A hazy supernate with suspended particles indicates an inadequate treatment, generally due to a low coagulant dose. Mostly, a 15-minute settlability test will determine the performance of the basin.

After coagulation and flocculation, the water flows to the sedimentation zone, which is the outer part of the basin. The flocculated water flows under or over a bonnet- (inverted cone) shaped baffle that separates the reaction and sedimentation zones. Solids settle to the bottom and are swept by the scrapers into one or more concentrating hoppers. A part of these solids is recirculated and the rest are discharged.

There are several different designs of solid contact basins by various companies that are suitable for different types of source water. Accelator and Walker systems are perhaps more popular and better known than the others.

Accelator Solid Contact Basin. In 1936, an accelator was first used in El Centro, California, to treat municipal water. In an accelator basin, incoming water and chemicals are mixed with the slurry of the previously formed floc in the primary mixing and reaction zone (lower central part of the basin) where coagulation takes place on the surface of the floc particles. Slurry concentration is kept at the desired level to provide the maximum amount of surface contact in this zone. To avoid the breaking up of the floc, a rotorimpeller (mixing unit) moves the flocculating water at a low velocity from the primary to the secondary reaction zone (upper central part of the basin). In the secondary reaction zone, floc formation is completed and the flocculated water is discharged outwardly and downwardly over an inverted conelike structure. This slurry forms a cloudlike suspension moving downward, with the lightest particles at the top and heaviest at the bottom. This cloud is called the *sludge blanket* (see Figure 6-3).

Clear water separates from the solids, which settle to the bottom. A part of these solids is drawn back to the primary reaction zone to maintain the 3 to 5 percent solid concentration. For proper operation, one sludge hopper is located close to the center, under the primary reaction zone, to collect the heavy solids, such as grit and sand. Two hoppers are located in the sedimentation zone close to the periphery of the basin. Heavy solids from the central hopper are discharged readily. Outer hoppers are used to discharge only the excess solids. To reduce the solid concentration, the sludge from the outer hoppers is discharged at a higher rate, and vice versa. Sludge is discharged by mechanical sludge valves operated at required intervals for the desired length of time. Clear water from the sedimentation zone passes through the launder (water collecting trough) holes below the water surface to avoid the carryover of floating solids.

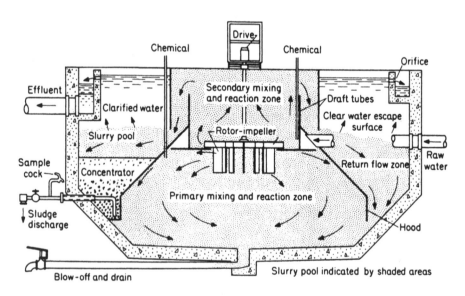

SOLIDS CONTACT UNIT

DRAWING REPRODUCED
BY PERMISSION OF "INFILCO"

Figure 6-3 Accelator Solid Contact Basin

(*Source: Journal of AWWA.*, 43(4) (April 1951) by permission. Copyright © 1951, American Water Works Association.)

Walker Solid Contact Basin. It has a coagulant added into the influent ahead of the unit. Coagulating water flows to the top central part of the mixing and reaction zone, where preformed floc is mixed and most of the chemical reactions for the coagulation are completed. The coagulated water flows through multiple tangential adjustable gates for uniform flow to the flocculation zone in an inverted cone-shaped compartment around the coagulation zone. For the proper functioning of the basin, these gates are properly adjusted for different flow rates. Flocculation is completed by slow mixing in the flocculation zone (see Figure 6-4).

Flocculated water flows uniformly outward under the cone to the sedimentation zone. After sedimentation, the clear water is collected from the top by overflow radial weirs, the overflow openings. Solids from the bottom are moved by scrapers to the central pit. A small part of these solids is recircu-

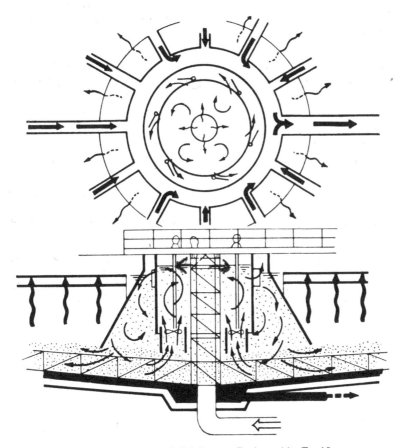

Figure 6-4 A Walker Solid Contact Basin and Its Top View
(*Source:* Walker Systems)

lated to the coagulation zone, and the rest is discharged by the sludge withdrawal system. Walker units are a little more sensitive to operate than accelators.

Both accelator and walker solid contact systems are termed *premix-recirculation* systems.

FINAL SEDIMENTATION BASIN

The final sedimentation basin is similar in structure to a conventional primary sedimentation basin. However, it has the provision for adding some chemicals, such as polymer into the water. This basin provides another step for turbidity removal by the further sedimentation of any carryover turbidity in the effluent of the primary sedimentation basin. Because there is either none or very little floc in the water, a small dose (0.5 mg/L) of a polymer is often used for coagulation of the turbidity. Generally, the final basin has a very small amount of sludge. Final basin effluent is, usually, disinfected with a small amount of chlorine for controlling the biological growth in the filter media. Water from these basins is crystal clear with turbidity less than 1 NTU. It flows to the filters for the final removal of turbidity.

FACTORS AFFECTING SEDIMENTATION

Several factors affect sedimentation:

- *Detention time.* An adequate detention time is important for complete sedimentation. The shorter the detention time, the less the settling, higher the turbidity, and vice versa.
- *Velocity.* Higher velocities cause the *scouring* (resuspension) of the settled floc, which may rise to the surface and cause high effluent turbidity. Very low velocity would not distribute the solids in the basin properly.
- *Surface turbulence.* The greater the turbulence, the less is the rate of sedimentation, and vice versa.
- *Short circuits.* Short-circuiting causes a short detention time, which results in an inefficient sedimentation.
- *Temperature.* The higher the temperature, the faster is the sedimentation, and vice versa. At higher temperatures water is lighter, and settling is faster.
- *Dimensions.* Proper dimensions of the basins are also important— particularly depth, which is generally 15 feet or more.
- *Inlets and outlets.* Inlets and outlets of the horizontal flow basins should be properly located to allow proper mixing of the water to prevent the short circuits.

Table 6-1 Sedimentation Problems and Their Solutions

Problems	Possible Causes	Possible Solutions
Poor sedimentation of a normal floc.	Sludge blanket is too high. This happens when fine sludge is not adequately removed.	Increase withdrawal of sludge until blanket is at least 5 feet below the water surface.
	Too much sludge in the basin. It could be due to less removal and more build-up.	Remove sludge until all heavy sludge is removed. Watch sludge density of discharged sludge. Removal should cease when sludge becomes thin and watery.
	Detention time is too short. Not enough time for complete sedimentation.	Check detention time and adjust, if possible.
	Short-circuiting of flocculated water; occurs in summer and winter due to stratification.	Check baffling and mixing mechanisms and increase the mixing.
	Inadequate flocculation due to excessive lime usage.	Reduce lime dose until sludge is denser.
Light feathery-flaky floc scouring into settled water.	Calcium carbonate precipitate happens during winter in waters with carbonate hardness and lime treatment in coagulation and flocculation phase.	Reduce lime dose and increase alum dose until floc becomes lighter.
Solids are gritty and like white sand.	Buildup of heavy solids in basin due to inefficient sludge removal.	Open sludge discharge valve; check sludge density; discharge until thin. Set timers to increase frequency and interval of discharge of sludge.
Scrapers of sedimentation basin are stopping.	Too much sand and silt in influent water.	Check presettled water turbidity and its nature. If turbidity is due to silt, then increase sludge removal.
	Some obstructing object such as a tool, piece of wood, or brick in basin.	Drain basin and look for obstructing object.

Table 6-1 (*Continued*)

Problems	Possible Causes	Possible Solutions
Clouds of floc are in the peripheral part of a solid contact basin.	If basin is a walker type, the tangential gates are not adjusted properly. If gates are too tight for a flow, they cause a higher velocity to the coagulated water. Floc cloud hits basin wall and moves upward to cause clouds.	Open gates until water level in coagulation zone is about two inches higher than that of the flocculation zone. Consult operational manual for proper adjustment of gates.
	In case of an accelator.	Slow down the central mixer speed. There might be insufficient sludge removal from outer sludge hoppers. Increase sludge removal from outer hoppers.
Slurry concentration is too low in solid contact basin.	In case of an accelerator-like basin.	Reduce sludge discharge from outer hoppers. Scrappers will bring in finer floc toward center for recirculation.
	In case of a walker-like basin.	Increase discharge frequency and reduce discharge interval to get rid of heavy solids and retain lighter solids for recirculation. If possible, increase speed of recirculation.
In solid contact basin slurry concentration is too high.	Opposite of previously defined problem.	Do opposite of previously defined problem. The more the sludge discharge of the fine floc, the less is the slurry concentration, and vice versa.
After heavy rain in summer, water does not flocculate and settle in solid contact basin.	High turbidity and low hardness.	This water can be treated at high pH with normal dose of coagulant (7–10 mg/L alum). May need small dose of a polymer. Sludge should be removed efficiently from basins. Sludge is heavy due to silt. Run jar test using a coagulant and a polymer; determine their optimum doses for best results and use the combination.

Symptom	Cause	Solution
There is poor sedimentation (in solid contact basin) of cold water with low turbidity.	Low pH.	High pH of the water will help in flocculation and sedimentation by providing OH⁻ ions to react with the coagulant. Add small dose of lime.
	Insufficient detention time. In winter, low temperature will increase the density of the water, which slows down sedimentation process.	Increase detention time by reducing flow through the basin. (Put on more basins.)
	Insufficient coagulant dose. Colder water needs higher dose of a coagulant for proper sedimentation.	Run jar test; determine required chemical doses and apply them.

- *Chemical feed points.* Feed points of the chemicals should be carefully located for proper reaction of each chemical with its target substance, e.g., alum should always be applied before lime.
- *Sludge withdrawal.* A proper withdrawal of the sludge is important to control the effective volume of the basin, the sludge decomposition, and the sludge scouring. Denser sludge settles readily and must be removed at a higher rate when compared to a lighter fluffy or flakey sludge.

SEDIMENTATION PROBLEMS AND THEIR POSSIBLE SOLUTIONS

Refer to Table 6-1 for some common sedimentation problems and possible solutions.

QUESTIONS

1. Define the term *sedimentation.*

2. What are short circuits?

3. Explain the term *solid contact process.*

4. Give some advantages of a solid contact basin as compared to a conventional basin.

7

SOFTENING

Softening of water is the removal of bivalent calcium and magnesium ions (Ca^{+2}, Mg^{+2}). These ions come from the dissolved compounds of calcium and magnesium. Their presence is known as *hardness* of water.

HARDNESS

Commonly, hardness is caused by the bicarbonates (HCO_3^-), sulfates (SO_4^{-2}), nitrates (NO_3^-), and chlorides (Cl^-), of calcium and magnesium. These minerals are dissolved by rain water and carried into the ground and surface waters. Areas with limestone $(CaCO_3)$ and dolomite $(CaCO_3$ and $MgCO_3)$ deposits have more hardness than others. Bicarbonates of calcium and magnesium are the most common forms of hardness. They are formed by the reaction of carbonic acid (H_2CO_3), which is carbon dioxide (CO_2) dissolved in rain water, with carbonate deposits.

$$CaCO_3 + H_2CO_3 \rightarrow Ca (HCO_3)_2$$

Hardness reacts with soap to form a grayish precipitate, the *curd*, rather than lather, and it makes the cleaning of clothes difficult. The curd formation continues until all hardness ions are used up. Therefore, the degree of hardness is judged as the measure of soap consumption to form lather. The harder the water, the greater consumption of soap is needed to do the cleaning, and vice versa.

Types of Hardness

- *Carbonate or temporary hardness* is mainly due to bicarbonates (HCO_3^-) of calcium and magnesium. It is called temporary because it can be removed by boiling the water, which converts some of the bicarbonates into insoluble carbonates.

$$Ca(HCO_3)_2 \rightarrow CaCO_3{\downarrow} + H_2O + CO_2{\uparrow}$$

- *Noncarbonate or permanent hardness* is caused by soluble compounds other than bicarbonates, such as sulfates, nitrates, and chlorides of calcium and magnesium. These compounds are more stable than bicarbonates; they are not removed by boiling the water. Calcium sulfate (gypsum) and magnesium sulfate (epsom) are the common causes of noncarbonate hardness.

Problems Caused by Hardness

Hardness is undesireable for several reasons. For example, hardness is

- A nuisance in laundering, due to wastage of soap and collection of dirty precipitate on fibers
- A nuisance in bathing
- A source of a dirty ring in the tubs and sinks.
- Responsible for a residue on washed objects like cars and utensils.
- Responsible for deposits on faucets and shower heads.
- Responsible for forming a carbonate scale inside the steam boilers.

SOFTENING

The water-softening methods can be classified as chemical precipitation and nonchemical precipitation methods.

Chemical Precipitation Methods

Lime or Lime–Soda Ash Softening Method. In this method, soluble calcium and magnesium compounds are converted to insoluble calcium carbonate ($CaCO_3$) and magnesium hydroxide ($Mg(OH)_2$), respectively. For this purpose, lime (quick lime, CaO, or slaked lime, $Ca(OH)_2$) and soda ash/ sodium carbonate (Na_2CO_3) are added to the water in the coagulation or flocculation basins. Lime is added after the alum, and soda ash is applied after the lime. This sequence is important for proper reactions. Alum needs to react with turbidity and precipitate it out before reacting with lime. After the alum and lime reaction, soda ash is added to react with permanent hardness (to prevent its reaction with alum or lime). *Lime and soda ash should never be fed through a common line because they will react and plug up the line.* Lime–soda ash reactions occur during the flocculation to form a part of the floc. Insoluble calcium carbonate and magnesium hydroxide settle out in the sedimentation basins along with the turbidity.

Lime removes all the carbonate hardness and noncarbonate magnesium hardness. It forms insoluble $CaCO_3$ and $Mg(OH)_2$ .The following chemical reactions show the removal of these hardnesses:

$$Ca(HCO_3)_2 + Ca(OH)_2 \rightarrow 2CaCO_3\downarrow + 2H_2O$$

$$Mg(HCO_3)_2 + 2Ca(OH)_2 \rightarrow Mg(OH)_2\downarrow + 2CaCO_3\downarrow + 2H_2O$$

$$MgSO_4 + Ca(OH)_2 \rightarrow Mg(OH)_2\downarrow + CaSO_4$$

These reactions show that removal of magnesium carbonate hardness requires twice the amount of lime than the removal of calcium carbonate hardness, and magnesium noncarbonate hardness removal produces the equivalent amount of calcium noncarbonate hardness, which needs soda ash for its removal. Soda ash removes the noncarbonate calcium hardness.

$$CaSO_4 + Na_2CO_3 \rightarrow CaCO_3\downarrow + Na_2SO_4$$

Removal of calcium chlorides and nitrates is similar, except that the products are sodium chloride and sodium nitrate. Due to a high amount of lime use, magnesium hardness removal needs pH above 10.6, whereas calcium is removed above pH 9.4.

These equations are used to determine the lime and soda ash doses corresponding to the degree of hardness removal.

Terminology of Lime and Lime–Soda Ash Softening Treatment Based on the Removal of Various Degrees of Hardness

- *Partial lime softening* uses a small amount of lime to remove the desired amount of calcium carbonate hardness.
- *Lime softening* is done with lime only. It is applied when water has only high carbonate hardness. This process requires pH of 9.6 to 9.8.
- *Excess lime softening* is the use of excess lime to remove high magnesium and calcium hardness. It needs pH above 10.6.
- *Lime–soda ash softening* is the use of lime and some soda ash. This process is used for waters with high calcium carbonate hardness, low magnesium hardness, and only some of the noncarbonate calcium hardness.

- *Excess lime–soda ash softening* uses excess lime and soda ash to remove high calcium and magnesium carbonate hardness and high noncarbonate hardness.

Dose Calculation. It is important to know how much of each chemical is needed to remove the desired amount of hardness. To calucale the hardness removal, determine the total alkalinity, total hardness, calcium hardness, and possibly the CO_2 contents of the water. As discussed before, total alkalinity is equal to carbonate hardness, and noncarbonate hardness is the difference between total hardness and alkalinity.

Lime also reacts with coagulants, carbon dioxide, iron, and manganese. Therefore, an excess amount of lime is needed. This can be determined by jar testing that takes into consideration all the reactants. The jar test simplifies the calculations.

Table 7-1 can be used to calculate the softening chemicals needed to remove hardness. For example, suppose, river water has 300 mg/L total hardness, 200 mg/L total alkalinity, and 200 mg/L calcium hardness. How much quick lime and soda ash will be required to remove 100 mg/L of carbonate hardness, 50 mg/L of noncarbonate calcium hardness, and 50 mg/L of magnesium carbonate hardness for lime-soda ash softening? All these values are in mg/L as calcium carbonate.

Carbonate hardness = Total alkalinity = 200 mg/L
Noncarbonate hardness = 300 mg/L − 200 mg/L = 100 mg/L
Calcium hardness = 200 mg/L

Thus, magnesium hardness = 300 mg/L − 200 mg/L = 100 mg/L

100 mg/L of calcium carbonate hardness needs,

(100×0.56) mg/L of CaO = 56 mg/L of CaO

50 mg/L magnesium carbonate hardness needs,

(50×1.12) mg/L = 56 mg/L of CaO.

Excess lime for magnesium removal and other reactants is an estimated 50 mg/L to raise the pH to 10.6.

Total quick lime required = (56 + 56 + 50) mg/L = 162 mg/L

50 mg/L of calcium noncarbonated needs,

(50×1.06) mg/L of Na_2CO_3 = 53 mg/L of $Na_2 CO_3$

Thus, we need 162 mg/L of quick lime and 53 mg/L of soda ash doses.

First, run a jar test on this water by using 162 mg/L of lime. After 10 to 15 minutes, add 53 mg/L of soda ash to the same sample and determine the

Table 7-1   Softening Chemicals Required to Remove Hardness

To reduce 1mg/L of:	Requires the following amount of (mg/L):		
	Quick Lime	Slacked Lime	Soda Ash
Noncarbonate hardness			1.06
Carbonate hardness, Ca	0.56	0.74	
Carbonate hardness, Mg	1.12	1.48	
Carbon dioxide	1.27	1.68	

Quick lime $=$ CaO

Slaked lime$=$ Ca (OH)$_2$

Calculated by stoichiometery from the softening equations

pH (which needs to be above 10.6) and actual hardness removal. Then adjust the dose as required. Apply the dose at plant-scale level and make the adjustments as needed to have the desired results.

It is cheaper to remove calcium hardness than magnesium hardness, and cheaper to remove carbonate hardness than noncarbonate; soda ash is a relatively expensive chemical when compared to lime.

Due to the lack of a required detention time and other unknown factors, it is difficult to produce waters with less than 50 mg/L hardness by lime–soda ash softening. In waters softened by this method, there is generally 50 to 80 mg/L total hardness with 30 to 50 mg/L of calcium. Mostly, municipal waters in the regions with high hardness are treated to have 100 $\pm$ 20 mg/L total hardness, which is cost effective and practical.

Lime–soda ash softening produces large amounts of solids. They are disposed off into a sludge lagoon, into a land fill, into the river, or onto the land.

Water pH, after lime or lime–soda ash softening, is high, and the water is very depositing. pH is adjusted by adding CO_2 into the final sedimentation basin. This process is known as *recarbonation*. pH is adjusted to below 9.3 to avoid any carryover calcium carbonate to precipitate in the filter media.

Nonprecipitation Methods

Membrane Softening. Membrane filtration, such as reverse osmosis, removes hardness without producing any residual solids. Currently, membrane filtration is used only for small operations. In the future, it may replace lime–soda ash softening treatment.

Ion Exchange Softening. This method uses zeolites (Z) or ion exchange resins that exchange calcium and magnesium ions for sodium or hydrogen ions, respectably. If hard water is allowed to stand in the sodium zeolite, Ca^{++} and Mg^{++} ions replace Na^+ ions and the water is softened.

$$CaSO_4 + Na_2Z \rightarrow CaZ + Na_2SO_4$$

An exhausted zeolite is regenerated by immersing it in a strong solution of sodium chloride (NaCl). Resins (R) act similar to zeolites. The reactions are as follows:

$$Ca(HCO_3)_2 + H_2R \rightarrow CaR + 2H_2O + 2CO_2$$
$$CaSO_4 + H_2R \rightarrow CaR + H_2SO_4$$

These resins are regenerated with acids. Water treated with a combination of cation and anion exchange resins is called *deionized* or *demineralized water.* Deionized water is used in analytical laboratories and bottling industries. This method is not practical for large municipal operations. Its use is limited to small industrial operations and individual water supplies.

Hardness removal is optional for a water utility; however, the public expects reasonably soft water. In natural water, the degree of hardness varies considerably in different regions. It varies from very soft (less than 10 mg/L) in most of the Eastern United States such as the Carolinas, to very hard (more than 300 mg/L) in Great Lakes region, like Michigan and Wisconsin. Midwestern States (e.g., Missouri and Kansas) have hard water (generally more than 300 mg/L). Soft, hard, and very hard water terminology is relative to each region and community. In the same metropolitan area, different water utilities may be treating water to different levels of hardness. For example, the Kansas City metropolitan area has several different levels of hardness in the treated water.

IRON AND MANGANESE IN WATER

The presence of other metal ions such as iron (Fe^{+2}) and manganese (Mn^{+2}) also cause hardness, but the amount is insignificant. They are commonly present in water with low pH and no dissolved oxygen. Both iron and manganese can be present in all groundwaters. Water containing more than 0.3 mg/L of iron and 0.05 mg/L of manganese is aesthetically objectionable. Soluble forms of iron and manganese are ferrous (Fe^{+2}) and manganous (Mn^{+2}) compounds. They are removed as insoluble ferric (Fe^{+3}) and manganic (Mn^{+3}) compounds.

Problems Caused by Iron and Manganese
- Iron and manganese stain clothes and enamel yellow and black, respectively.
- They are undesirable for bottling, laundries, paper mills, tanning, and ice manufacturing.

- Iron deposits in the distribution systems cause red water complaints and inaccurate meter readings.

These metals can be removed by precipitation and nonprecipitation methods similar to calcium and magnesium removal. In all the precipitation methods, they form ferric oxide (Fe_2O_3) and manganic oxide (Mn_2O_3). Ferric oxide is commonly called rust.

Chemical Precipitation

- *Lime–soda ash treatment* removes iron and manganese while removing calcium and magnesium. They are removed above pH 9.4. They precipitate out as Fe_2O_3 and Mn_2O_3. When hydrated (wet), they are ferric hydroxide ($Fe(OH)_3$) and manganic hydroxide ($Mn(OH)_3$).
- *Aeration* is the adding of oxygen into water. It is done by passing air through the water in aeration towers, by cascading, or mechanical aerating. Oxygen reacts with iron and manganese compounds to form ferric and manganic oxides. For effective iron removal, pH should be around 7.5. Manganese oxidizes slower than iron and requires a higher pH.
- *Chlorination* is the process by which chlorine also removes iron and manganese.
- *Chlorine dioxide and ozone treatment* use the disinfectants chlorine dioxide and ozone to remove iron and manganese.

Nonprecipitation Method

The zeolite treatment is a nonprecipitation treatment. Removal of iron and manganese by the cation exchanger method is similar to and simultaneous with the removal of calcium and magnesium. Water should not be aerated before the zeolite treatment to avoid any accumulation of ferric hydroxide in the zeolite bed.

SOFTENING PROBLEMS AND POSSIBLE SOLUTIONS

Refer to Table 7-2 for some common problems and possible solutions.

Table 7-2 Softening Problems and Their Solutions

Problems	Possible Causes	Possible Solutions
Soda ash does not effectively remove hardness.	Hardness is gone up in the raw water.	Run hardness test; determine right dose by jar test; apply it.
	Lime and soda ash react together before they react with the hardness.	Check feed lines because they might be feeding too closely. Feed lime first and soda ash later.
	Feeder does not feed correctly.	Check feeder belt speed; look for any obstruction in feeding system, and correct it.
Soda ash line plugs up.	Mixing of lime slurry with soda ash solution.	Check lines and any other possible mixing of these two chemicals. They react and produce calcium carbonate. Calcium carbonate will deposit in the lines.
Soda ash solution is milky and gritty.	In some way, there is a mixing of some lime with soda ash in the storage bin.	Check soda ash bin for any contamination with lime.
Lime is not effectively softening.	Lime dose is too high. It will be indicated by the light and flakey floc and slightly milky water due to undissolved calcium hydroxide which causes high calcium hardness.	Check lime dose by jar testing and reduce it as required.
	Coagulant (alum) dose is too low.	Run jar test; determine right dose of coagulant; and feed correct dose. Alum will react with lime and reduce hardness.
	Quick lime (CaO) is not slaking properly. It is indicated by grit still slaking in the grit drum, which means an improper grade of lime.	Ask supplier to provide proper grade of lime.
During winter, softening sludge is heavy, gritty, and like white sand.	An insufficient alum dose. Especially during winter, an improper dose of alum can cause sandlike calcium carbonate sludge formation that separates from the rest of the lighter sludge. It is easily visible as a bottom layer in a graduated cylinder while running the settlability test. If not removed, this sludge can lock up scrapers (as discussed previously).	Run jar test and determine optimum alum dose until all the sludge has a uniform density. It does not stratify in the jar test.

QUESTIONS

1. Softening of water is the removal of calcium and magnesium. T or F
2. Explain the following terms:
   a. Temporary hardness
   b. Permanent hardness
   c. Partial lime softening
   d. Recarbonation
3. Give the sequence of application of lime, soda ash, and alum into the treatment train.
4. State the pH levels for removal of calcium and magnesium. Which one is less expensive to remove?
5. At pH 9.6, calcium, iron, and manganese all precipitate out. T or F
6. Identify various problems caused by high hardness, iron, and manganese.
7. State three methods of removing hardness and discuss their application in the water utilities. Which of these three is most commonly used in municipal water treatment, and why?

8

STABILIZATION

Stabilization is treating water to be neither corrosive nor depositing. Lime–soda ash softened water with a high pH and high calcium content could be highly depositing (scaling), and soft coagulated water with low pH could be highly corroding. Commonly, the goal of proper water treatment is to produce water that is slightly depositing, to protect the distribution system and the plumbing fixtures from corrosion. It also protects the public health from corrosion byproducts such as lead and copper. The levels of these metals in the drinking water are controlled under the lead and copper rule of 1991. Stabilization protects the system from too much deposition or too much deterioration.

STABILIZATION OF DEPOSITING WATER

Lime–soda ash softened water, generally, has high calcium content and can cause an excessive calcium carbonate scaling of basins, lines, filter media, and plumbing fixtures. If not stabilized, it may form a very thick (1/2–1 inch or thicker) layer in the pipes that will limit their capacity and even plug up some small pipes. It can cause a serious loss of water flow and pressure. Calcium carbonate scaling is controlled by lowering the pH of the softened water in the final sedimentation basin by *recarbonation* (passing of carbon dioxide gas through the water). Normally, lime–soda ash softened water has pH 9.5 or higher, depending on the degree of softening. Generally, it is adjusted to 9.3 to keep the excess calcium carbonate in solution. Softened water may also need sequestering to control the limit of calcium carbonate deposition.

Sequestering

Sequestering keeps calcium in solution. For this purpose, a small dose of a sequestering agent, such as sodium hexametaphosphate, commonly known as *Calgon* (which means calcium gone) is used. Besides controlling the excessive calcium carbonate scale formation, it also dissolves rust, which reduces the number of red water complaints. Being highly corrosive, sodium hexametaphosphate corrodes iron in the absence of calcium. Therefore, its dose should be carefully controlled. Generally, after recarbonation, 0.5 mg/L dose is applied to the filter influent to prevent excessive scaling of the media. Another small dose is applied to the filter effluent to control the scaling in the distribution system. As a rule of thumb, the total dose should be less than 1 mg/L, and at the farthest end of distribution, there should only be a nominal amount of residual sodium hexametaphosphate.

Indicators

The degree of calcium carbonate deposition is determined by several indicators and indexes. Most of them are based on the calcium content, alkalinity, and the pH of water.

- *Observations.* Visual observations of the pipe sections, pipe coupons (part of the pipe cut out for making water connections), primary and final sedimentation basins, and filter media are important to note the extent and nature of calcium carbonate deposition. Record the age, thickness, color, and the nature of roughness of the scaling of different pipes. We should keep a good record to see the past effect and the future trend to determine the required changes in the water treatment.
- *Marble test.* This test is based on the alkalinity of the water and its calcium carbonate deposition trend. The water to be tested is saturated with calcium carbonate ($CaCO_3$) by adding $CaCO_3$ powder to it, shaking, and keeping it overnight. Alkalinity of the sample is determined before and after the saturation. The difference between the initial and the final alkalinity measure determines the nature of water. The difference values of more than 0, equal to 0, and less than 0 indicate depositing, stable, and corrosive water, respectively. It is a simple, easy, and good test for an operator to determine the water's stability. If a marble is kept overnight in this saturated water, it will show a layer of calcium carbonate if the water is depositing.
- *Baylis Curve.* This curve shows the calcium carbonate equilibrium at different pH and alkalinity values (see Figure 8-1). The point of intersection of these values determines the quality of the water. The intersection point above the curve, on the curve, or below the curve indicates depositing, stable, and corrosive water, respectively. It is a very useful

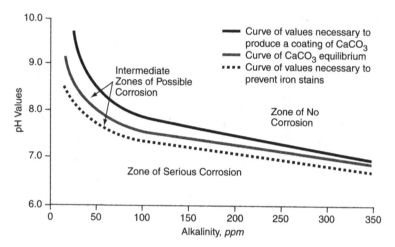

Figure 8-1 Baylis Curve Showing Relationship between pH and Alkalinity Values (*Source: Basic Chemistry for Water and Wastewater Operators,* by permission. Copyright © 2002, American Water Works Association.)

curve because in the water treatment pH and alkalinity are the commonly adjustable parameters for stabilization.

- *Langlier Saturation Index* (*LSI*). This is the most commonly used index in water treatment (see Figure 8-2). Besides alkalinity and calcium contents of the water, it requires temperature, total dissolved solids, and pH to determine the $CaCO_3$ deposition capacity of water. The LSI is based on the pH of the water at the calcium carbonate saturation point, which is called *pHs*. The difference of the actual pH and pHs determines the nature of a water. If this difference is 0, above 0, or below 0, water is stable, scaling, and corrosive, respectively. The pHs value is determined from the following nomogram.

$$LSI = pH - pHs$$

- *Ryzner Index* (*RI*) *or Stability Index.* This is a modification of LSI for a more precise measure. It uses the same parameters. RI is the difference of 2 pHs and pH of the water.

$$RI = 2\ pHs - pH$$

An RI value of less than 6 indicates depositing water, and more than 6 indicates corrosive water. The higher the RI, the more corrosive is the water.

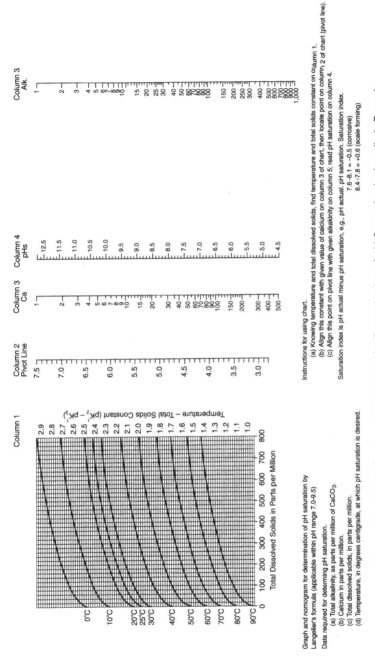

Figure 8-2 Riehl's Graph and Nomogram for Determination of pH Saturation by Langlier's Formula

(*Source: Basic Chemistry for Water and Wastewater Operators*, by permission. Copyright © 2002, American Water Works Association.)

STABILIZATION OF CORROSIVE WATER

Corrosive water corrodes structures, lines, and plumbing fixtures. It is stabilized by adjusting its pH and alkalinity by adding lime, sodium hydroxide (lye), sodium carbonate, or sodium bicarbonate to bring the pH and alkalinity above the corroding level.

Principles of Corrosion

Corrosion is the dissolving of metals like iron, lead, copper, cadmium, zinc, tin, and antimony from the structures, pipes, plumbing fixtures, and solders. It is natural to return the processed metals into their natural dissolved state. It is a complex process because a large number of environmental variables are involved. A generally accepted explanation is that it is an electrochemical reaction that is formed of three parts: anode, cathode, and an electrolyte solution. The *anode,* the positive electrode, is the site where the reducing agent, a metal (iron, lead, or copper), is getting dissolved by losing electrons. The *cathode,* the negative electrode, is where an oxidizing agent (oxygen or H^+) accepts the electrons. The *electrolyte solution,* the conducting medium, is the water with dissolved electrolytes. Completion of reaction needs all these three parts. It is like an electric circuit.

Here is an example of the corrosion of iron in water:

Anode (positive electrode)

$$Fe \rightarrow Fe^{+2} + 2e^-$$

These two electrons are accepted by an oxidizing agent like O_2 or an acid (H^+) at the cathode.

Cathode (negative electrode)

$$1/2\ O_2 + H_2O + 2e^- \rightarrow 2(OH^-)$$

Overall reaction with oxygen as an acceptor is

$$Fe^{+2} + 2(OH^-) \rightarrow Fe(OH)_2$$

or

$$2Fe + O_2 + 2H_2O \rightarrow 2Fe(OH)_2$$

Ferrous hydroxide, $Fe(OH)_2$, is grayish, which is further oxidized into rusty, insoluble ferric hydroxide, $Fe(OH)_3$, in the presence of dissolved oxygen:

$$4\ Fe(OH)_2 + O_2 + 2H_2O \rightarrow 4Fe(OH)_3\text{, Rusty slurry}$$

Oxygen is present in natural waters as dissolved oxygen (DO). Under acidic conditions, H^+ is the acceptor:

$$2H^+ + 2e^- \rightarrow H_2\uparrow$$

The H^+ ion is contributed by any substance that acts as or produces acid in the water. These substances are carbon dioxide, hydrogen sulfide, chlorine, alum, chlorides, sulfates, nitrates, and water itself due to its self ionization (discussed in the pH test).

$$H_2O \leftrightarrows H^+ + OH^-$$

Overall reactions with acid and water are

$$Fe + 2HCl \rightarrow FeCl_2 + H_2\uparrow$$

$$Fe + 2H_2O \rightarrow Fe\ (OH)_2 + H_2\uparrow$$

Therefore, natural water with dissolved substances has all three required parts for corrosion to occur; thus, almost any metal in contact with water will corrode. The rate of corrosion will depend on the nature of metal and water characteristics.

Other metals, such as lead and copper, are corroded the same way as iron and are mostly converted into their oxides.

Types of Corrosion

These are some of the important classes of corrosion based on its physical, chemical, and biological nature:

- *Physical corrosion* is the erosion of a pipe surface due to a high velocity (of over 5 ft/sec), particulate matter, or dispersing gas bubbles.
- *Galvanic/bimetallic corrosion* occurs when two dissimilar metals are connected together in the water lines, such as lead and copper, lead solder on copper line, and a galvanized iron fitting connected to a copper pipe. In the electromotive series, elements are arranged according to their oxidizing and reducing strength. Cesium, the most active reducing agent (metal), is at the top. Mercury, the least active, is at the lowest level of the metals. Position of a metal in the series determines whether it will corrode. A metal at a higher level in the electromotive series becomes an anode and will corrode, while the one below it becomes the cathode (see Table 8-1). This principle is applied in the cathodic protection

Table 8-1 Electromotive Series

Reduced State	Oxidized State	Oxidation Potential, *volts*
Cs	$Cs^+ + e^-$	3.02
Li	$Li^+ + e^-$	3.02
K	$K^+ + e^-$	2.99
Ba	$Ba^{++} + 2e^-$	2.90
Ca	$Ca^{++} + 2e^-$	2.87
Na	$Na^+ + e^-$	2.71
Mg	$Mg^{++} + 2e^-$	2.34
Al	$Al^{+3} + 3e^-$	1.67
Mn	$Mn^{++} + 2e^-$	1.05
Zn	$Zn^{++} + 2e^-$	0.76
Cr	$Cr^{+3} + 3e^-$	0.71
Fe	$Fe^{++} + 2e^-$	0.44
Co	$Co^{++} + 2e^-$	0.28
Ni	$Ni^{++} + 2e^-$	0.25
Sn	$Sn^{++} + 2e^-$	0.14
Pb	$Pb^{++} + 2e^-$	0.13
H_2	$2H^+ + 2e^-$	0.00
Sn^{++}	$Sn^{+4} + 2e^-$	−0.15
Cu	$Cu^{++} + 2e^-$	−0.34
$2I^-$	$I_2 + 2e^-$	−0.53
Fe^{++}	$Fe^3 + e^-$	−0.75
Hg	$Hg^+ + e^-$	−0.80
Ag	$Ag^+ + e^-$	−0.80
Hg	$Hg^{++} + 2e^-$	−0.85
Hg^+	$Hg^{++} + e^-$	−0.91
$2Br^-$	$Br_2 + 2e^-$	−1.06
$2Cl^-$	$Cl_2 + 2e^-$	−1.36
$2F^-$	$F_2 + 2e^-$	−3.03

Left axis: Increasing Strength of Reducing Agent

Right axis: Increasing Strength of Oxidizing Agent

method of the corrosion control. In case of lead solder on a copper pipe, lead acts as an anode and copper as a cathode, causing the corrosion of lead. Similarly, lead will dissolve when lead and copper lines are connected together.

- *Stray current corrosion* occurs due to grounding of appliances through the pipes. Corrosion occurs where the stray current leaves the pipe. It is the cause of mysterious red water complaints in some homes. Grounding of appliances should never be through the water lines.

- *Localized or pitting corrosion* starts when an area of an otherwise coated metal surface is exposed due to faulty coating, damage, or stress. This area becomes anodic and forms a pit, and the surrounding area serves as a cathode. As the corrosion proceeds, the pit grows and has blackish-gray ferrous hydroxide inside and an insoluble rusty, ferric hydroxide layer on the outside. This outside growth is known as a *tubercle*. High velocity, flushing against the normal flow direction in the lines, or water hammer can rupture these tubercles and cause the red water complaints. Pitting can seriously damage the water lines.

- *Bacterial corrosion* is caused by bacteria, such as iron bacteria, nitrogen bacteria, and sulfur bacteria, which become active under low dissolved oxygen and low disinfectant conditions. These conditions normally prevail at the dead ends of pipes, causing corrosion, slime formation, high heterotrophic plate counts, and sometimes even the presence of coliform bacteria. Water quality deteriorates and then the water, generally, has a bad smell. When velocity in the water mains increases, some of this smelly water reaches homes and causes taste and odor complaints. Generally, water utilities have a regular unidirectional flushing plan to resolve this problem.

Factors Affecting Corrosion

- *pH.* This is one of the most important parameters in stabilizing the water. As a rule, the lower the pH, the higher the acidity, and higher is the rate of corrosion. In case of scaling water, it is lowered close to 9; and in case of corrosive waters, it is raised above 7. Treated water pH should never be lower than 7.

- *Alkalinity.* Along with pH, alkalinity determines the degree of stabilization of water. The higher the alkalinity, the higher is the deposition of calcium carbonate, and less corrosive is the water. A higher alkalinity requires a lower pH for stabilization, and vice versa.

- *Hardness.* Generally, the higher the hardness, the higher is the calcium carbonate content, and lower is the rate of corrosion.

- *Temperature.* Higher temperature causes faster corrosion reactions and a higher corrosion rate. There is less corrosion during winter than during summer.

- *Dissolved gases.* Naturally dissolved gases in the water are oxygen and carbon dioxide. Both provide cathodic conditions. The higher the amount of these gases, the higher is the rate of corrosion.

- *Polarization and depolarization.* During the corrosion process, when the cathode surface is covered with hydroxides (Fe $(OH)_2$) and H_2 molecules, it creates a barrier for the further reactions and controls the corrosion.

This physical barrier is known as *polarization*. Removal of this barrier is called *depolarization,* which results in the resumption of corrosion. Depolarization can be caused by the increasing acidity, which neutralizes hydroxides; by increasing oxygen concentration, which combines with hydrogen to form water; or by increasing the water velocity which sweeps away hydroxides and hydrogen gas.

Corrosion Control

Almost any metal in contact with water will corrode to some extent, depending on the environmental conditions. *In the water treatment, the corrosion process can be slowed but not stopped completely unless none of the metal part is exposed to water.* A full protection needs a protective barrier between water and the metal. This barrier is provided either by the manufacturer of the pipes as a coating—such as cement lining with the tar coating, paint, plastic, or rubber—or provided by the treated water as a coating of calcium carbonate, a phosphate, or a silicate.

Protective Coatings Provided by the Treated Water

- *Calcium carbonate coating* is the most commonly used due to the presence of natural hardness in the water. $CaCO_3$ coating size and rate of deposition depend, mainly, upon the pH, alkalinity, and the calcium carbonate content of the water.
- *Phosphate coating* is provided by the three types of phosphates: orthophosphates, polyphosphates, and zinc phosphates. *They form a protective layer on the cathodic site* by reacting with corrosion products and metal.
  - *Orthophosphates* are simple phosphate compounds, such as phosphoric acid (H_3PO_4), sodium phosphate (Na_3PO_4), sodium monohydrogen phosphate (Na_2HPO_4), and sodium dihydrogen phosphate (NaH_2PO_4).
  - *Polyphosphates* are long-chain phosphates formed by reacting phosphoric acid with sodium or potassium compounds. A common example of a polyphosphate is sodium hexametaphosphate, $(NaPO_3)_6$. In the presence of calcium and iron ions at a low pH, they form a protective coating on the cathodic site. At high pH and low dose, they dissolve iron and calcium by a sequestering mechanism, thus preventing excessive scale formation. For corrosion control, they require high velocity of water and a high dose. Polyphosphates remove corrosion products from the anode, form positively charged colloidal particles, and deposit them on the cathode area. Increasing polarization of the cathode reduces corrosion.
  - *Zinc phosphates* contain zinc in various concentrations (10 to 30 percent) with ortho- or polyphosphates. The protective film is formed of zinc compounds. The higher zinc concentration acts faster by rapid film formation. The higher the pH, the lower the zinc content is required for an effective control.

- *Silicate coating* has been used by some water utilities in the eastern United States and Canada. Sodium silicate ($Na_2 SiO_3$) is the most commonly used chemical. It is available as a dry or liquid product. Silicates are used for waters with pH 7–9. For proper corrosion control, a 4 to 30 mg/L dose is required. Silicates form a protective thin layer of their compounds over the anode, the corroded metal part; therefore, unlike phosphates, they are anodic inhibitors. This film, unlike calcium carbonate, does not become very thick. If the application is stopped, the film breaks down and protection stops. Silicates can be combined with zinc or with phosphates for better corrosion control by protecting the anode and the cathode. Silicates are more expensive than phosphates; therefore, their use is not very common.

Cathodic Protection Cathodic protection is a noncoating corrosion control method. It is reversing the corrosion process by providing a more active anodic metal than the metal to be protected, which becomes cathodic. The anodic metal corrodes and the cathodic metal is protected. Cathodic protection is used to protect basins and their parts. Anodes of aluminum are suspended in a basin; their length and spacing are determined by the water characteristics and the area of the tank. Electric circuits are installed in the metal to be protected. A mild electric current is passed through these circuits.

The basic theory is to provide a current from an outside source through the anode to the tank. All surfaces of the tank should get uniform current for a successful treatment. Anode dissolves, so it should be checked regularly and replaced as needed.

Problems Caused by Corrosion

- The structures, lines, water heaters, and plumbing fixtures are damaged.
- The taste and odor become evident.
- Corrosion products shelter microbes from disinfectants that cause an after growth and slime formation in the distribution system.
- Higher level of disinfectant is demanded in the distribution system.
- Fixtures and laundry are stained.

Refer to the Table 8-2 for some common stabilization problems.

Table 8-2 Stabilization Problems and Their Solutions

Problems	Possible Causes	Possible Solutions
Corrosion of some parts of the basin with the excessive build-up of calcium carbonate.	Some unprotected parts of metal are exposed to water.	All parts of basin should be protected by paint, cathodic protection or both.
	Ineffective cathodic protection.	Check the anodes. If they are used up, replace them.
		Check the electrical parts of the circuit & correct them, if needed.
Excessive deposition in the pipes; pH is 9.	Water alkalinity is above 100 mg/L. Rate of calcium carbonate deposition depends on alkalinity and pH of the water. A higher pH needs a lower alkalinity for the proper rate of deposition, and vice versa.	See the Baylis Curve and lower the pH (by recarbonation) for the proper stability.
	Langlier Saturation Index is too high. The higher the index, the more is the deposition.	Lower the index to around $+0.3$ plus or -0.1.
Sodium hexametaphosphate causes corrosion of lines.	Low calcium content of the water. Sodium hexametaphosphate is a very corrosive substance. In the absence of calcium carbonate it will attack iron.	Check calcium hardness and lower the sodium hexametaphosphate dose until there is a small amount of its residual in the water.
	Feeder overfeeds.	Check feeder and setting and correct it.
Treated water is of depositing quality, but there is corrosion in the distribution system.	Some parts of distribution system (like fittings) are unlined. These exposed parts will serve as anodes and will corrode.	Check for exposed parts and replace them with lined ones.
	There may be a grounding of household appliances through the water lines. The problem will be mainly within the house's plumbing.	Check for such a situation and remove grounding wire from the water line.
	Connection between two dissimilar metals causes galvanic corrosion.	Check for such connections (like lead and copper or a galvanized iron fitting and copper) and replace them with proper metals.

QUESTIONS

1. Explain these terms:
   a. Stabilization
   b. Recarbonation
   c. Sequestering
   d. Polarization
   e. Depolarization

2. How are depositing and corroding water stabilized?

3. For stabilization, which two parameters are commonly adjusted in the water treatment?

4. Explain the Baylis Curve and Langlier Saturation Index and their application in stabilization.

5. If the Langlier Index of water is -0.5, is the water corrosive or depositing?

6. Name four parameters used to calculate the Langlier Index.

7. What kind of health problems are caused by corrosion?

8. What is the purpose of the lead and copper rule?

9. Corrosion is an electrochemical reaction. T or F

10. Explain the functioning of the corrosion process.

11. If lead and copper are connected together in the water lines, which metal will dissolve?

12. Most unprotected metals will corrode when exposed to water. T or F.

13. Name some protective coatings that are mechanically applied by the manufacturer of a pipe.

14. Name three types of coatings that can be provided by the treated water.

9

FILTRATION

Filtration is the mechanical removal of turbidity particles by passing the water through a porous medium, which is either a granular bed or a membrane. Filtration's purpose is to remove all the turbidity particles carried over from the sedimentation phase, thus producing a sparkling clear water with almost *zero* turbidity.

GRANULAR MEDIA FILTRATION

A granular media filter, generally, consists of a rectangular concrete structure with 4-feet-deep media formed of sand or a combination of sand, garnet, anthracite (crushed hard coal), and activated carbon (see Figure 9-1). The media are supported by a layer of gravel. Under the gravel is a drain system for the drainage of filter effluent, called *filtrate*. Mostly, a small amount of cationic polymer is applied to the filter influent for micro flocculation. Polymer and turbidity particles form a very fine floc that accumulates on the top of the filter media and forms a straining mat (also called a *surface cake*) that removes the turbidity. Turbidity is removed by two mechanisms, straining and adsorption. Adsorption is acquiring the turbidity particles on the surface of micro floc. Most of the turbidity is removed in the top few inches of media (see Figure 9-2).

There is a slightly high turbidity during the first 10 to 15 minutes of the filtration because the mat is not effectively formed. This is known as the *ripening period,* after which filtration is adequate. When there is too much build-up of the surface mat and filter interstices are plugged up, the rate of filtration decreases, and turbidity starts going up. At this point, the filter needs backwashing.

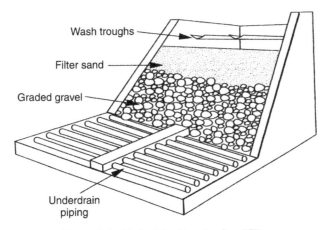

Figure 9-1   Vertical Section of a Sand Filter
(*Source: Water Treatment: Principles and Practices of Water Supply Operations,* published by American Water Works Association, 1995.)

Backwashing is the removal of filtered-out turbidity by reversing the flow through the filter (i.e., from the bottom upward). The time period from beginning filtration to the filter wash is called a *filter run*. The period from the start of filtration to the end of the backwashing is called a *filter cycle*. Turbidity of filter effluent and the resistance to flow, called *head loss,* are monitored continuously to determine the backwashing time and the filter

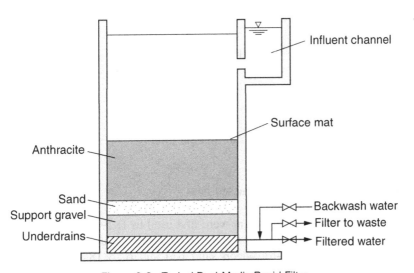

Figure 9-2   Typical Dual-Media Rapid Filter
(*Source:* Adapted from MWH, *Water Treatment, 2e.* Copyright © 2005 by John Wiley & Sons, Inc. Reprinted by permission of John Wiley & Sons, Inc.)

performance. Generally, a washed filter is taken out of service for at least 30 minutes for the proper settling of media before putting it back into operation.

A good filter operation removes more than 99 percent of the feed water turbidity and produces a sparkling clear water with turbidity as low as 0.1NTU or less.

Particle Size and Density

Particle size and density of a granular medium is expressed by three parameters: uniformity coefficient, effective size, and specific gravity. The first two parameters are determined by sieving a sample of medium through a set of standard sieves with pore size as millimeters (mm). Two sieves are selected, one that allows 60 percent of the media to pass through and retains 40 percent, and a second one that allows 10 percent of the media to pass through and retains 90 percent.

Uniformity coefficient is the ratio of the pore size of the first sieve to the second. *Effective size* is the pore size of the second sieve. If the pore size that allows 60 percent of a medium to pass through is 0.75 mm and the pore size of the sieve that allows only 10 percent to pass through is 0.45 mm, then the uniformity coefficient of this medium is 0.75 mm/0.45 mm = 1.66, and the effective size is 0.45 mm. *Specific gravity* is the ratio of the density of the medium to the density of water. It determines the vertical stratification of different media in the filter bed, with the lightest at the top and the heaviest at the bottom.

Types of Granular Filters Based on Media, Filtration Rate, or Principle of Operation

The following list shows the types of filters:

- Slow sand filters
- Rapid sand filters
- High-rate sand filters
- Granular activated carbon multimedia filters
- Pressure filters

Each type of filter will be discussed next.

Slow Sand Filters. These filters were first used in 1829 to treat the London, England, water supply. A slow sand filter is a covered underground concrete structure with a 3- to 5-foot-deep sand bed and 6 to 18 inches of graded gravel, which has the largest size at the bottom and the smallest at the top (see Figure 9-3). Effective size of sand particles is 0.25 to 0.35 mm, with the

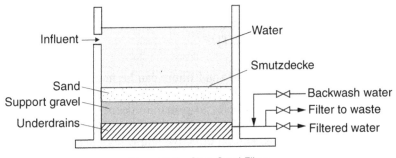

Figure 9-3 Slow Sand Filter

(*Source:* Adapted from MWH, *Water Treatment, 2e.* Copyright © 2005 by John Wiley & Sons, Inc. Reprinted by permission of John Wiley & Sons, Inc.)

uniformity coefficient 2.5 to 3.5. Media are supported by the under drain system. The filter cover is at least 6 feet above the media. The filter is operated with 3- to 5-feet-deep water above the medium. Water flows slowly through the medium and leaves most of the turbidity particles in the top layer. Loading rate is 0.03 to 0.06 gallon per minute per square foot (gpm/ft²) of the filter surface.

Turbidity particles form a surface mat that becomes sticky due to microbial activity. This mat is called *smutzdecke,* which is very effective to remove particles by straining, adsorption, and microbial metabolism. After the filter run, which could be several days or even weeks, the filter is taken out of service and cleaned. For cleaning, the top layer of sand is scraped, washed, and stored for replacement. The filter is cleaned several times by scraping the surface layer before replacing any sand. For an effective filtration, the minimum required depth of sand is 2 to 2.5 feet. There is no backwashing in these filters. These filters are effectively used for direct filtration of source water with very low (less than 1 NTU) turbidity such as pristine mountain streams or reservoirs.

Rapid Sand Filters. Unlike the slow sand filters, surface loading in these filters is 2 to 4 gpm/ft.², and there is backwashing after the filter run. Sand depth, in these filters, is 2 to 3 feet. The particles have an effective size of 0.35 to 0.55 mm and uniformity coefficient of 1.6. Medium is supported on 18 inches of gravel, which is graded from 4 inches to pea size. The under drain system has a Leopold- or Wheeler-type false bottom for an effective drainage of the filter effluent. To facilitate the uniform flow of the water, the Leopold system has blocks with small holes, and the Wheeler system has conical rectangular cavities with balls. During filtration, there are about 30 inches of water above the medium. *Free board,* the distance between the surface of the medium and the lip of the backwash trough, is 24 to 27 inches to prevent any loss of medium during the backwashing. Filtration takes place in the top few inches of the medium. These filters are used to filter water with influent turbidity up to 5 NTU.

High-Rate Sand Filters. Rapid sand filters can be modified to create high-rate sand filters. A coarser and lighter layer of anthracite is applied above the sand to allow the turbidity particles to penetrate deeper into the media. Due to deeper penetration of particulate matter, these filters allow a higher rate of filtration, longer filter runs, and an effective and economical filtration. These filters are operated at 5 to 10 gpm/ft.² loading. They have two or three media stratified according to their size, shape, and specific gravity. The lightest and coarsest medium is at the top, and the finest and heaviest medium is at the bottom. There are two types of high-rate filters: dual media and multimedia filters.

- *Dual media filters* have two media, which are anthracite and sand. Generally, the filter bed from top to bottom is formed of 18 to 30 inches of anthracite, 12 inches of sand, and 12 inches of gravel. Anthracite, the crushed hard coal with angular particles and larger voids between particles, is lighter than sand. The effective size of anthracite is 0.6 to 0.7 mm, the uniformity coefficient is 1.6, and specific gravity is 1.55. Sand has rounded particles, which are more compacted with smaller voids. Effective size, uniformity coefficient, and specific gravity of sand are 0.45 to 0.5 mm, 1.5 to 1.7, and 2.65, respectively. These specific gravities keep anthracite and sand well stratified after backwashing. Anthracite and sand trap the larger and smaller turbidity particles, respectively. These filters are quite common and popular among most of the water systems.
- *Triple media/mixed media filters* are a modification of dual media filters. A third layer of the heaviest medium is applied under the sand. Mostly, this layer is garnet, which is heavier and finer than sand. Garnet has effective size of 0.2 to 0.3 mm and specific gravity of 4.2. From top to bottom, a typical triple media filter has 36 inches of anthracite, 18 inches of sand, 8 inches of garnet, and 8 inches of gravel. Garnet removes the smallest turbidity particles. There is some mixing of the media at the interface of adjacent layers, which makes them *mixed media filters.*

Granulated Activated Carbon (GAC) Multimedia Filters. GAC filters have a layer of activated carbon on top of anthracite or sand. Activated carbon adsorbs various contaminants, such as tastes and odor-causing organics, THMs, and synthetic organics. GAC is lighter than sand or anthracite and has an effective size of 0.55 to 0.65 mm with a uniform coefficient of 2.4. These filters have the problem of losing some carbon during the backwashing; therefore, backwashing is properly controlled to prevent the excessive loss of GAC. Commonly, backwashing causes 1 to 6 percent GAC loss per year.

All granular media filters discussed to this point are gravity flow filters.

Pressure Filters. In these filters, media are enclosed in a cylindrical steel tank and the water is forced under pressure through the filter. Media are either sand or diatomaceous earth.

Rapid sand pressure filter has an 18 to 24-inches-thick sand layer with gravel underneath. The filter rate is 2–5 gpm/ft.^{2} Being small, their use is limited to some industries and recirculation of swimming pool water.

Diatomaceous earth pressure filters have diatomaceous earth medium. Diatomaceous earth is a light medium formed of commercially available diatom fossils with particle size of 5 to 50 micrometers (μm). As compared to several inches of sand, thickness of this medium is only 0.06 to 0.12 inches. Turbidity particles are retained on the surface, and there is hardly any penetration of them into the medium. Generally, the filtration rate is 1 gpm/ft.^{2} These filters are used by the small water systems for low turbidity source water.

Filter Backwashing

There is no standard criterion for backwashing of a filter. Mostly, it is decided by the performance of the filter from effluent turbidly, head loss, and filter run. For example, turbidity should not be more than 0.1NTU, head loss should not be more than 6 feet (pressure as water height in feet), and filter run no longer than 24 hours. These are general guidelines, which vary from plant to plant.

Filter Backwashing Procedure. Following is a general step-by-step procedure for the manual filter wash:

1. Close the influent valve and let the water level drop to about 4 to 6 inches above the medium.
2. Close the effluent valve.
3. Gradually, start the surface wash system, which will loosen the surface mat of suspended material. Surface washing is done by revolving jets, by compressed air scrubbing, or by mechanical rakes.
4. Open the backwash water valve gradually to prevent the media waste.
5. Open the wastewater drain valve. Wash until wash water is quite clean. Proper cleaning may take up to 10 minutes.
6. Stop the surface wash at least 2 minutes before closing the wash water valve.
7. Close the wastewater drain valve.

8. Let the media stratify properly.

9. To put the filter back in service, open the influent valve and then the effluent valve.

Modify this general procedure as required for a particular utility.

For uniformity of washing, a large number of plants have an automatic filter backwash system, which works fine; it needs to be monitored to make the necessary changes in its programming.

Proper backwashing requires about 50 percent expansion of the fluidized (suspended) sand. The percentage of *sand expansion* is calculated by using a stick with small panes at different heights. The stick is placed on top of the media while washing the filter. The highest pane that gets some sand is the point to which sand is expanded. The percentage of expansion is the percent of sand particles rising. It is the rise of sand divided by the depth of the sand media and then multiplied by 100. For example, if the rise is 15 inches and the depth of media is 30 inches, the expansion is (15 in./30 in.) × 100) = 50 percent.

Backwash volume should not be more than 2.5 percent of the total water filtered.

Factors Affecting Granular Media Filtration

- *Turbidity.* The less the turbidity in the filter influent, longer the filter run, and better is the performance.
- *Media form.* The coarser the media the less is the head loss, the longer is the run, and vice versa.
- *Depth.* The deeper the bed, the better is the filtration.
- *Backwashing.* Proper backwashing is an important factor in the proper operation of a filter. Improper washing can cause the loss of media, mixing of media, formation of mud balls, cracks, and craters. All these factors cause an inadequate filtration and a high-effluent turbidity.
- *Filtration rate.* The higher the loading, the shorter the filter runs, and less efficient is the filter.
- *Temperature.* The higher the temperature, the better is the performance.
- *Water stability.* In the lime softening plants, higher pH (above 9.3) and higher calcium carbonate content of water can cause deposition of calcium carbonate on the media particles. This build-up of calcium carbonate causes swelling of media and the formation of mud balls. Water needs to be stabilized by lowering the pH below 9.3. A controlled small amount of a polyphosphate, such as sodium hexametaphosphate, is applied as a sequestering agent to further correct this situation. Too much of a polyphosphate can cause excessive sloughing of calcium carbonate from the media particles, which causes higher turbidity, and too little may not be enough for an adequate sequestering.

- *Polymer dose.* A small dose (0.5–0.75 mg/L) of a polymer is helpful in forming a micro-floc mat to aid the filtration. A higher dose causes cracks in the filter mat, and a lower dose does not form an effective microfloc.

MEMBRANE FILTRATION

This process is the passing of pretreated water under pressure through a membrane to remove specific sized particles. A *membrane* is a very thin paperlike structure. Membranes can achieve the degree of treatment comparable to a conventional treatment plant. Membrane treatment is one of the best treatment technologies to meet the present and expected SDWA challenges. It is capable of removing most of the regulated contaminants.

Membrane Structure

Membranes are either hollow fine fiber (HFF) or spiral wound (SW) structures formed of cellulose acetate and synthetic materials, such as polypropylene or poly furon.

Hollow fine fiber membrane is a hollow fine hairlike tubular structure called *fiber*. The fiber is folded in a U form. A bundle of thousands of these U-like fibers is packed in a tubular pressure vessel, which increases the filtering surface tremendously. These pressure vessels, the tubes, make the hollow fiber membrane system very compact and efficient. Water flows through the membrane leaving concentrated water, called *concentrate* (retentate or waste) on the influent side and filtered water, called *permeate* on the other side of the membrane. Flow-through the membrane can run inside out or outside in. They are called *cross flow* and *transverse flow,* respectively. A cross-flow membrane has concentrate inside and permeate outside, while a transverse-flow membrane is just the opposite. Hollow fine fiber is a popular and commonly used form of membranes in water systems (see Figure 9-4).

Spiral wound membrane is a flat sheet rolled around a central tube, which collects permeate (see Figure 9-5). The sheet has an outside active membrane and a thick fibrous support or backing underneath. Usually, two membranes are placed back to back with a separator in between. They are rolled together to form a membrane element. These elements are housed in a cylindrical pressure vessel.

Mechanism of Particle Removal

There are three basic mechanisms: sieving, selective diffusion, and charge repulsion:

1. *Sieving.* Each membrane has a uniform and specific pore size. All particles bigger than the pore size are sieved out/rejected. The smallest

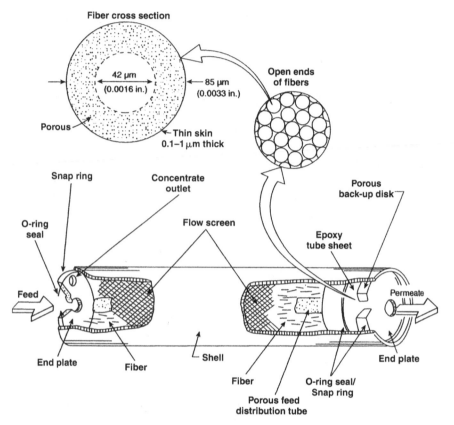

Figure 9-4 Hollow Fine Fiber Membrane

(*Source: M-46—Reverse Osmosis and Nanofiltration.* American Water Works Association, 1999.)

rejected particle is slightly bigger than the pore size. The smallest molecular weight that will be strained out is known as a *cut-off molecular weight* (COMW).

2. *Selective diffusion.* Passing only selected dissolved particles through the membrane is selective diffusion. For selective diffusion, a membrane needs to be semipermeable, which means it will allow only certain chemicals to diffuse through, and all others will be rejected (e.g., reverse osmosis membranes).

3. *Charge repulsion.* Filtration uses a direct electric current; anions go to the anode, and cations go to the cathode, e.g., electrodialysis membranes. Electrodialysis is dialysis aided by electrodes. *Dialysis* is the separation of dissolved and suspended substances by a membrane.

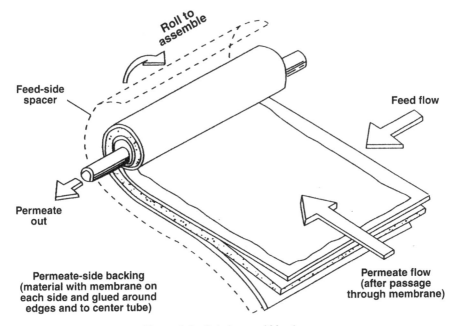

Figure 9-5   Spiral-wound Membrane

(*Source: M-46—Reverse Osmosis and Nanofiltration.* American Water Works Association, 1999.)

Types of Membranes

Based on the pore size and mechanism of functioning, membranes can be divided into five groups:

1. *Microfiltration membranes* function like a sieve. Their pore size ranges from 0.1 to 1 micrometer (μm); therefore, they remove all particles bigger than 1 μm, including *Cryptosporidium* oocysts, *Giardia* cysts and all bacteria. They are successfully used for water treatment plants with less than 12 million gallons per day (mgd) capacity and low raw water turbidity. For example, the 5 mgd Saratoga, California, water treatment plant uses micro filtration membranes with 0.2 μm pore size.

2. *Ultrafiltration membranes* are similar to the microfiltration membranes except the pore size is 0.003 to 0.1 μm to remove very small particles. They remove all particles bigger than this pore size, including viruses and THM formation precursors. They remove cysts and other pathogens by six logs, meaning 99.9999 percent removal. They are more expensive due to the smaller pore size. The finer the pore size, the more effective the membrane, and the more expensive it is.

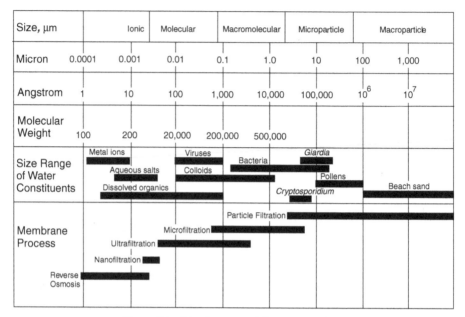

Figure 9-6   Membranes and Contaminant Size

(*Source:* HDR Engineering, Inc. *Handbook of Public Water Systems, Second Edition.* Courtesy of Dept. of Interior, Bureau of Reclamation, Water Quality Improvement Center, Yuma, AZ.)

3. *Nanofiltration membranes* have nanometer (0.001 μm, or 1 nm) pore size. They remove all the particles above nanometer size. Besides removing viruses, cysts, and bacteria, they remove some dissolved substances.

4. *Reverse osmosis membranes* are semipermeable. They function as sieves and selective diffusion membranes due to osmosis, which allows some specific dissolved substances to pass through. *Osmosis* is the passage of water through a semipermeable membrane from the lower concentration of the dissolved substances to the higher concentration to equalize the concentration on both sides of the membrane. The force with which water flows through the membrane is called *osmotic pressure.* The greater the difference in concentration on two sides of the membrane, the higher the osmotic pressure, and faster is the flow. In *reverse osmosis,* a pressure higher than the osmotic pressure is applied on the higher concentration side to force the water through the membrane in the reverse order.

These membranes will remove all the suspended particles larger than the pore size and only selective dissolved substances. These membranes remove substances like sodium, calcium, magnesium, and other metal compounds. They are used to treat seawater that has total dissolved solids (TDS) in the range of 3.5 percent (35,000 mg/L) and other brackish water (TDS up to 3 percent).

Reverse osmosis has been used since 1960 for desalting brackish water. In the United States, the first municipal brackish water treatment plant was built in 1971 at Ocean Reef Club, Florida, to treat 0.6 mgd flow. At present, there are more than 100 reverse osmosis drinking water plants in the United States. It is a common process in Middle Eastern countries to treat salty water for drinking. Reverse osmosis is also used by the carwash industry to reduce the total dissolved solids in the municipal water supplies.

5. *Electrodialysis membranes* use direct electric current to separate dissolved electrolytes from the water. Anions are collected at the anode and cations at the cathode, after passing through a resin membrane. Unlike all membrane systems, they are not pressure driven.

Membrane Fouling

Membrane fouling is the clogging of the membrane by the filtered-out matter, which forms a layer called *cake* on the membrane surface. The degree of fouling depends on the quantity of particulate matter in the feed water and the level of its removal. The fouling is determined by applying indexes such as the *silt density index* (SDI) and the *modified SDI*. To determine SDI, two portions of 500 mL of the feed water are passed through a 0.45 μm pore size and 47 mm diameter membrane filter under 30 psig (pounds per square inch gauge pressure) of constant pressure. The time period is recorded (in minutes) for the initial and final 500 mL of filtration as t_i and t_f, respectively. Final filtration is generally after 5, 10, or 15 minutes from the start of the initial filtration, depending on the quality of the feed water. This period is the third reading called the total time, t_t. SDI is calculated by the following formula:

$$\text{SDI} = [100\,(1 - (t_i/t_f)]/t_t$$

where:

t_i = time to collect initial 500 mL
t_f = time to collect final 500 mL
t_t = total time, mostly 15 minutes as the maximum

Example. Suppose it took 2 minutes to collect the initial 500 mL; and after 15 minutes from the start of the test, it took 3 minutes to collect the final 500 mL.

$$\text{SDI} = 100\,(1 - (2\ \text{min}/3\ \text{min}))/15\ \text{min} = 2.22$$

Feed water should have an SDI value of less than 5. Follow the manufacturers' specifications for specific systems.

With the proper pretreatment, it may take weeks or even months before membrane cleaning is required. The cleaning is done by using compressed air along with backwashing to remove the debris, or chemical cleaning with warm (100°F) sodium hydroxide and a surfactant mix.

Membrane Integrity

Membrane integrity is the soundness of the membrane condition regarding leak damage. It is determined by various direct and indirect tests. There are important direct and indirect tests that are used for this purpose.

Direct Tests

- *Bubble point.* A 29.4 psi pressure is applied with a dilute surfactant (soap) solution. A 0.7 psi pressure decrease in 5 minutes indicates a damaged membrane, which will also be indicated by bubble formation.
- *Air pressure hold test.* Pressure decrease is checked after 10 minutes. Perhaps, it is the most reliable direct test.
- *Sonic sensor.* An online device indicates any unusual sound in the system.

Indirect Tests

- *Effluent turbidity.* Any abrupt increase in effluent turbidity indicates a leak in the membrane system.
- *Flux rate.* Flux is the rate of water flow through the membrane expressed in gallons per day per square feet (gpd/ft.^{2}). A sudden increase in flux indicates a damaged membrane.
- *Particle count.* Any sudden increase in particle count indicates a leak in the system.

Membrane Filtration Treatment Plant

A membrane filtration plant differs from a conventional treatment plant in being compact. Essentially, it is formed of three phases: pretreatment, membrane process, and posttreatment (see Figure 9-7).

Pretreatment. *Pretreatment* is the preparation of feed water so that there is a minimum of membrane fouling. The main pretreatment methods are microstraining and conventional treatment up to final sedimentation.

- *Microstraining.* Raw water is filtered with 5–13 μm pore size strainers without using any chemicals. This is a common practice for surface water with low turbidities.
- *Conventional treatment up to final sedimentation.* Conventional treatment is used for turbid surface water with natural organic matter (NOM).

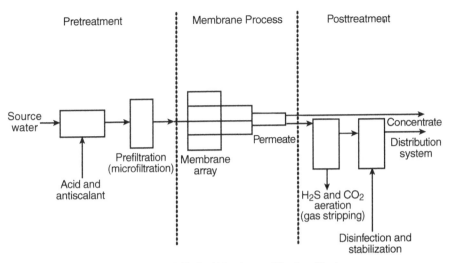

Figure 9-7 A Typical Membrane Filtration Plant

(*Source: M-46—Reverse Osmosis and Nanofiltration.* American Water Works Association, 1999.)

Membrane Treatment. Membrane filtration uses an appropriate membrane system that takes into account source water quality and treatment requirements. Currently, there are three commonly used membrane systems: microfiltration, ultrafiltration, and reverse osmosis. The first two are the most feasible alternatives to conventional water treatment to meet the requirements of the enhanced surface water treatment, the total coliform, and the disinfectants and disinfection byproducts rules. Here are some guidelines to select a membrane system for different source water:

- *Microfiltration* is used for surface water where conventional treatment requires pretreatment, sedimentation, and filtration.
- *Ultrafiltration* is used for surface water that needs the removal of very small particles such as viruses and dissolved organics.
- *Reverse osmosis* is suitable when water has a very high concentration of dissolved substances such as chlorides, nitrates, and fluorides, in addition to other contaminants (e.g., salty water).

Posttreatment. In posttreatment, the effluent is treated to make sure that filtered water is safe and stable. Commonly, it needs aeration, pH adjustment, and postdisinfection.

- *Aeration.* To remove carbon dioxide and hydrogen sulfide, aeration is used.

Table 9-1 Filtration Problems and Their Solutions

Problems	Possible Causes	Possible Solutions
Mud balls are in a sand filter.	High level of CaCO$_3$ in water. Calcium carbonate forms aggregates in the surface mat.	Lower the pH (below 9.3) of the influent water to convert CaCO$_3$ into soluble calcium bicarbonate. Also add 0.5–0.75 mg/L sodium hexametaphosphate to the filter influent to keep calcium in solution.
	Abrupt high rate of washing. Mud balls are formed by faulty distribution of washwater caused by abrupt and quick opening of the washwater valve. It causes openings in the filter bed by pushing the media and letting some of the surface mat aggregates of CaCO$_3$ fall below the media's surface. They grow close to 1″ in size and can cause clogging of filter and high effluent turbidity.	Start the filter washing by surface agitation and gradually increasing the wash water flow.
Cracks are in surface mat of a sand filter.	High dose of polymer to filter influent. Micro floc forms the gelatinous surface mat. When the mat becomes thick, it cracks at weak points due to water pressure. Cracks form water channels to let water pass through the media without effective filtration. Polymer feed for micro flocculation should be lowest effective dose—generally 0.25 to 0.75 mg/L.	Determine required polymer dose by jar test and decrease it as required.
	Longer filter runs can cause a thicker mat which can crack at thinner places.	Shorten the filter run adequately.
There is jet action and sand boiling in a mixed media filter.	Abrupt opening of the backwash water valve. Fast movement of backwash water in part of the filter bed can cause boiling of media particles resulting in media mixing and poor filtration.	Open wash water valve slowly and partially for first few minutes of the washing.

Problem	Cause	Solution
There is high turbidity in filter effluent of a conventional treatment plant when all other phases have no problems.	High pH of the filter influent: pH above 9.4 causes fine calcium carbonate ($CaCO_3$) particles to go through filters to cause a high turbidity reading.	Lower pH to 9.3 which will convert calcium carbonate to soluble calcium bicarbonate. Remember there is no insoluble calcium carbonate below pH 9.3.
	Overfeed of a polyphosphate: A dose of a polyphosphate such as sodium hexametaphosphate over 0.75 mg/L to the filter will cause excessive removal of calcium carbonate coating of the filter media, which results in a high turbidity reading.	Run a jar test; determine optimum dose of polyphosphate, and apply. Generally, 0.5 mg/L dose is adequate.
	Overfeed of a polymer. Overdose of a polymer to the filter influent will cause cracks and channels in the media for the turbidity particles to go through without filtration.	Check polymer dose to the filter influent. Usually, a dose above 0.5 mg/L can cause a high turbidity reading. Run jar test; determine optimum polymer dose and apply.
	Air bubbles in the turbidity measuring cell. There is a higher concentration of dissolved gases in the water, especially in winter, which start coming out as gas bubbles at room temperature.	Warm the sample to remove gases before taking the turbidity reading.
	Scratched or smudged sample cell.	Always use a scratch-free sample cell since scratches will also give a false high turbidity reading. Wipe the cell with soft tissue paper.

Table 9-1 *(Continued)*

Problems	Possible Causes	Possible Solutions
Surface sweeps are being buried under the media.	High calcium carbonate content. High calcium carbonate deposition on media particles will cause the swelling of media.	Lower the pH below 9.3 and apply small amount of polyphosphate as previously discussed. The amount of swelling can be determined by washing a sample of media with 5 percent hydrochloric acid. The difference of the dried sample weight or volume before and after washing, times 100, is the percent of swelling in weight and volume, respectively. It may require the acid washing of the whole media to reduce its volume. This will also improve the filter performance.
There is biofouling of a membrane.	High bacterial count in feed water.	Check bacterial count in pretreated water and reduce it by proper predisinfection.
	Membrane has been out of service too long. Some bacteria in and on a wet membrane will start to multiply if the membrane is out of service too long.	Flush membrane with filtered water and sanitize it as recommended by the manufacturer.
There is chemical scaling of the membrane.	A high metallic (calcium, magnesium, and iron) salt content. These metals will become concentrated in the feed water during filtration and precipitate.	Use an antiscalant such as sodium hexametaphosphate or polyacrylate as recommended by the manufacturer.
	pH of feed water is too high. This will cause more deposition of metallic salts.	Lower the pH to dissolve the scaling substances.

- *pH adjustment.* Generally, effluent pH is low with high corrosion potential. It is adjusted by using sodium hydroxide. Sodium hydroxide also converts carbon dioxide in the effluent to bicarbonate alkalinity.

$$NaOH + CO_2 \rightarrow NaHCO_3$$

- *Postdisinfection.* There is a possibility of contamination of the effluent with any leak of feed water. Normally, chlorine is used for postdisinfection.

Waste Disposal

Concentrate forms 20 to 50 percent of the feed water. It is very high in dissolved substances and considered as a wastewater or industrial waste; therefore, its disposal is regulated under the Clean Water Act, the Safe Drinking Water Act, and the Underground Injection Regulation. Concentrate is disposed into surface water, into sewer system, into deep wells, or into evaporation ponds.

Membrane filtration is a good option for small systems such as those with less than 15 mgd capacity and good-quality source water. This technology is still not cost-effective for large utilities. Furthermore, it is much more sophisticated and high tech operation than a conventional treatment plant.

Filtration is the final conventional treatment step to remove turbidity and thus the waterborne pathogens.

FILTRATION PROBLEMS AND THEIR POSSIBLE SOLUTIONS

Refer to Table 9-1 for some frequent filtration problems.

QUESTIONS

1. What is the main purpose of filtration? Name two filtration systems.

2. Name the two most commonly used granular filter media in the conventional water treatment plants.

3. Explain these terms:
   a. Effective size
   b. Uniformity coefficient
   c. Specific gravity
   d. Head loss
   e. Smutzdecke

4. a. Write the major differences between slow sand and a high-rate sand filter.

    b. What is the source of diatomaceous earth?

    c. Why is a granulated activated carbon cap used on a dual media filter?

5. a. If there is a mixed-media filter formed of garnet, anthracite, and sand, what will their configuration be in the filter bed, and why?

    b. Multimedia or mixed media filters mean the same thing. T or F

6. a. Identify various devices for surface washing of granular media filters.

    b. Backwashing should start at a low flow and gradually increase to the required washing flow. T or F

    c. Write two criteria for backwashing. State the commonly used head loss for the backwashing.

7. Write the basic steps in the filter backwashing.

8. Explain these terms:

    a. Mud balls

    b. Cracking

    c. Sequestering agent

9. a. State three mechanisms of contaminant removal in membrane processes.

    b. Which one of the following two systems removes more particulate matter, microfiltration or reverse osmosis?

10. Describe briefly:

    a. Microfiltration

    b. Ultrafiltration

    c. Reverse osmosis

    d. Nanofiltration

11. Define these terms:

    a. Spiral wound membrane

    b. Hollow fine fiber membrane

    c. Fouling

    d. Membrane integrity

    e. Silica density index

    f. Permeate

    g. Concentrate

12. State a direct and an indirect test to determine the integrity of a membrane.

10

DISINFECTION

Disinfection is the destruction or inactivation of waterborne pathogens. Most of the waterborne pathogens enter the source water through sewage, and the rest enter through the watershed runoffs. They originate from human and animal wastes, respectively. Human wastes carry the pathogens causing cholera, typhoid, paratyphoid, bacillary dysentery, amebic dysentery, and infectious hepatitis. Animal wastes carry *Cryptosporidium parvum,* causing cryptosporidiosis, coming from the cow and *Giardia lamblia,* causing giardiasis, coming from the beaver.

The last two pathogens are hard-to-kill opportunistic human parasites. Until 1989 when the surface water treatment rule was promulgated, the control of *Giardia lamblia* was the most serious problem; after the 1993 Milwaukee, Wisconsin, cryptosporidiosis episode, *Crytosporidium parvum* became the most difficult pathogen to control. Perhaps that episode was the reason for enactment of the enhanced surface water treatment rule that emphasized the further removal of turbidity and adequate disinfection. Just as filtration is aimed at the complete removal of turbidity, disinfection of filtered water, called *postdisinfection* (disinfection after treatment), is aimed at the complete destruction of waterborne pathogens.

The removal and killing of waterborne pathogens have been the targets in water treatment since pretreatment. Predisinfection, disinfection during the treatment, sedimentation, filtration, and postdisinfection serve as steps to remove, kill, or inactivate waterborne pathogens. Postdisinfection purposes to disinfect filtered water and leave an adequate disinfectant residual in the treated water to ensure its safety in the distribution system until its use by the farthest consumer.

Perhaps, water disinfection was started with the boiling of water by Hippocrates about 400 B.C. to make it safe. The old practice of two-minutes brisk boiling of water to ensure safety is still practiced in some countries for individual water supply and is recommended in the United States by water utilities when the water treatment breaks down.

Jersey City, New York, and Bubbly Creek, Chicago, were the first public utilities to disinfect their water with chlorine in 1908. After that, chlorine has been the common disinfectant in the United States and other countries. In 1974, the reactions of chlorine with natural organic matter (NOM) to form harmful THMs were discovered. After this discovery, alternative disinfectants such as chloramines, chlorine dioxide, ozone, and ultraviolet light radiation were studied and considered. Subsequently, it was found that most chemical disinfectants form harmful byproducts. Chlorine forms THMs and haloacetic acids (HAA); chloramines form cyanogen chloride and chloropicrin; chlorine dioxide forms chlorites and chlorates; and ozone forms bromates and aldehydes. These byproducts are controlled under the disinfectants and disinfection byproducts (D/DBP) rule. Thus, each utility requires a careful disinfection plan to have a good balance between proper disinfection and disinfection byproducts. This balance is the goal of each utility (see Figure 10-1).

Achieving balance is a challenge for water utilities. Utilities with source water with very low THMs potential, low turbidity, and pH range 7–8 may use free-residual chlorine successfully. However, those with high THMs potential, variable pH, and high turbidity use chlorine dioxide or ozone for predisinfection, followed by free-residual chlorine or/chloramines for postdisinfection. Some utilities use or consider using UV light radiation treatment.

METHODS OF DISINFECTION

Chlorination

The application of chlorine is still the most common method of disinfection. It is economical, effective, and helpful in controlling tastes and odors, iron and manganese, slime-producing bacteria, cyanides, and phenols.

Properties of Chlorine. Chlorine is a greenish-yellow gas at room temperature with the boiling point, melting point, and density as $-34.6°C$, $-101°C$, and 3.21 g/L, respectively. It can easily be compressed into liquid; 450 mL of gas forms 1 mL of liquid chlorine. It is 2.5 times heavier than air. It irritates the nose and burns skin. It is a very strong oxidant reacting with various organic substances, ammonia, and metals. Chlorine inactivates microorganisms by reacting with their enzymes.

In water, chlorine produces hydrochloric acid and hypochlorous acid, which lower the water pH. Hypochlorous acid, a strong disinfectant, is the principal disinfecting form of chlorine. Any chemical that produces hypochlorous acid in water is considered a form of chlorine.

GOAL

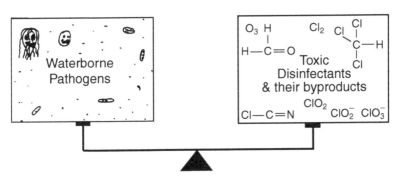

Figure 10-1 Goal of Disinfection of Water

Water + Chlorine → Hydrochloric acid + Hypochlorous acid

$$H_2O + Cl_2 \rightarrow HCl + HOCl$$

At pH above 7.5, hypochlorous acid ionizes progressively into hypochlorite ion (OCl⁻) and hydrogen ion (H⁺); thus, it becomes less and less effective. The higher the pH, the higher is the ionization, and less effective the chlorine.

$$HOCl + H_2O \rightarrow H_3O^+ + OCl^-$$

Forms of Chlorine

Pure Chlorine. Compressed, pure, and liquid chlorine is available in 100-, 150-, and 2,000-pound containers, and tank car lots with 15 to 90 tons of capacity. It is applied to water by means of a special apparatus called a *chlorinator.*

Hypochlorites. Hypochlorites of calcium and sodium are solid forms of chlorine. They react with water to produce hypochlorous acid and hydroxides. Unlike chlorine gas, they raise the pH of the water. High-test hypochlorite, commonly known as *HTH*, is calcium hypochlorite (Ca (ClO)$_2$), which contains 70 percent available chlorine. Sodium hypochlorite (NaClO), the common household bleach (e.g., Clorox), solution contains 15 percent available chlorine. Hypochlorite solution is applied by *hypochlorinators*. All hypochlorites are corrosive to some degree. Thus, they are stored in wood, glass, plastic, or rubber containers.

Calcium hypochlorite + Water → Calcium hydroxide + Hypochlorous acid

$$Ca\ (ClO)_2 + 2H_2O \rightarrow Ca(OH)_2 + 2HOCl$$

$$NaClO + H_2O \rightarrow NaOH + HOCl$$

HTH is used for sterilizing water lines and disinfecting swimming pools, and chlorine bleach is used for municipal water disinfection. Chlorination by using hypochlorites is called *hypochlorination.* Use of chlorine gas for water treatment is difficult for some utilities due to safety requirements that necessitate the use of chlorine bleach. An old and concentrated bleach may contain the contaminant chlorate, which has health effects; thus, a fresh and diluted bleach solution is used.

Breakpoint Chlorination. This chlorination technique ensures proper disinfection of water containing iron, manganese, organic matter, and ammonia by producing free-residual chlorine. The purpose of breakpoint chlorination is to produce and maintain free-residual chlorine in the water after complete oxidation of substances that react with chlorine.

Increasing doses of chlorine are added to a series of samples; after reaction time the chlorine residual is measured and plotted against the dose, which results in a distinct curve called *breakpoint curve.* This curve gives a much better understanding of chlorine reactions in water and the disinfecting form of chlorine at different stages of chlorination (see Figure 10-2).

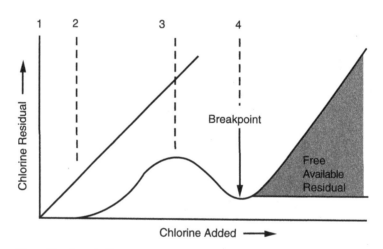

NOTE: Numbers indicate points within treatment where chlorine is added.

Figure 10-2 Break Point Chlorination

(Source: Basic Chemistry for Water and Wastewater Operators. Copyright © 2002, American Water Works Association. Reprinted by permission.)

At first, there is no residual; then the residual increases to the highest point, *hump.* Afterward, a further increase in dose results in a decrease of residual, until the lowest point, the *breakpoint* or *dip,* is reached. After the breakpoint, an increase in dose results in a corresponding increase in residual. This residual chlorine is known as *free available* or *free residual chlorine* or *breakpoint residual,* which is hypochlorous acid under normal pH (below 7.5). The breakpoint indicates the completion of all the reactions of chlorine with all the reactants in the water. Breakpoint chlorination, therefore, ensures a proper disinfection after control of iron, manganese, bacteria, tastes, and odors. Reactions occur in the following sequence:

1. *Reducing agents.* Iron, manganese, hydrogen sulfide, and nitrites neutralize chlorine into chlorides. Thus, iron, manganese, and hydrogen sulfide—which cause red water, black water, tastes and odors, respectively—are removed. There is no residual chlorine at this stage.
2. *Organic matter.* Organic substances including microorganisms react with chlorine to produce chloroorganic compounds, which show as residual chlorine and can cause taste and odor problems.
3. *Ammonia.* Chlorine reacts with ammonia and forms chloroamines until the hump.

$$\text{Ammonia} + \text{Hypochlorous} \rightarrow \text{Monochloramine} + \text{Water}$$

$$NH_3 + HOCl \rightarrow NH_2Cl + H_2O$$

Both chloroorganic compounds and chloramines form *combined residual chlorine.*

4. *Destruction of combined residual chlorine.* Chlorine reacts with chloroorganic compounds and chloramines and neutralizes them, resulting in a drop of residual chlorine. These reactions occur between the hump and breakpoint. After the breakpoint, chlorine is free-residual chlorine, generally as hypochlorous acid.

Chlorine demand is the amount of chlorine consumed in the water during the contact period. It is the difference between the dose and residual chlorine after the contact time. For example, assume we used 3 mg/L chlorine dose in a sample, and after 30 minutes there was 2 mg/L residual. Chlorine demand of this water is 3 mg/L − 2 mg/L, or 1 mg/L.

Factors Affecting Chlorination

- *pH.* Effectiveness of chlorine is pH dependent. The lower the pH, the more effective the free-residual chlorine, and vice versa. Both hypochlorous acid and hypochlorite ions are called free-residual chlorine. The amount of hypochlorous acid, the principal disinfectant, decreases at above pH 7.5 as it ionizes into hypochlorite ions. The following table suggests the minimum free residual chlorine required at different pH values for disinfection.

pH	Free Residual Chlorine
6–8	0.2 ppm
8–9	0.4 ppm
9–10	0.8 ppm

These amounts at 20°C will disinfect water satisfactorily in about 10 minutes.

- *Residual chlorine type.* Hypochlorous acid is more effective than the hypochlorite ions.
- *Temperature.* Effectiveness of chlorine varies directly with the temperature. The higher the temperature, the quicker the disinfection and shorter is the required contact time.
- *Contact time period.* Chlorine requires a certain amount of contact time at different temperatures to react with microorganisms. The longer the contact time, the more effective is the disinfection. Thus, the SDWA adopted the *CT* concept to ensure a proper disinfection. *CT* stands for concentration of a disinfectant as mg/L and its contact time in minutes. Required *CT value* is a constant for each disinfectant at different temperatures. In chlorination, it also takes into consideration the pH of the water. The lower the pH, the better the disinfection, and lower is the required CT value.
- *Concentration.* The higher the concentration, the more effective is the disinfectant. Normally, 0.5 to 1 ppm free-residual chlorine will effectively disinfect the water. According to the SDWA, the maximum allowed chlorine residual in the distribution system is 4 mg/L and the minimum required is 0.2 mg/L.

Chlorine and Health. Chlorine is a very toxic gas, even in small concentrations in air. The Occupational Safety and Health Act (OSHA) allows less than 1 ppm of chlorine in the air. The following chart shows the physiological effects of various concentrations of chlorine by volume in the air.

Effects	Chlorine as ppm in Air
Least detectable by odor	below 3.5
Produces throat irritation	15
Produces coughing	30
Dangerous for 30 minutes exposure	40–60
Rapidly fatal	1,000

Chlorine and Safety. Observe the following precautions when dealing with chlorine:

- *Use a mask when entering a chlorine-containing atmosphere.* Mask should be kept outside the chlorine room and checked regularly for leaks.
- *Check chlorinator, lines, and cylinder valves regularly for leaks.* Use ammonia fumes to test leaks. Ammonia and chlorine combine to produce white fumes of ammonium chloride, indicating a leak.
- *Always store chlorine on the lowest floor because it is heavier than air and collects at the lowest level.* Never stoop down when there is a chlorine smell in the room. Observe and meet all the OSHA requirements.

Chloramination

Chloramination is the use of *chloramines* for disinfection. Chloramines are produced by reacting ammonia with chlorine; they are slower and less effective than the free-residual chlorine (HOCl). Therefore, they require a higher dose and longer contact time than the free-residual chlorine for the same degree of disinfection. Despite these drawbacks, they have been used successfully by a large number of utilities. Chloramination was first used in the United States in 1917 at Denver, Colorado. Ammonia is used ahead of chlorine to prevent THM formation.

There are three species of chloramines: monochloramines, dichloramines, and trichloramines. Type of chloramine formation depends on the pH and chlorine-to-ammonia ratio. Above pH 8 and ratio of chlorine to ammonia of 4 to 1, monochloramine is the predominant species, which is preferred for disinfection because it is quite effective and less odorous. Below pH 8 and the progression of chlorine to ammonia ratio, the species changes from monochloramines to dichloramines and from dichloramines to trichloramines, as shown by the following reactions.

$$NH_3 + HOCl \rightarrow NH_2Cl + H_2O$$

$$NH_3 + 2HOCl \rightarrow NHCl_2 + 2H_2O$$

$$NH_3 + 3HOCl \rightarrow NCl_3 + 3H_2O$$

Chloramination is more effective above pH 8. Its effectiveness is comparable to hypochlorite ions. Its advantages are that there is no THM formation and it has a longer residual effect than chlorine. It is commonly used in postdisinfection for a longer residual effect in the distribution system.

Besides slow and weak disinfectants, chloramines induce hemolytic anemia and cause problems with the use of dialysis machines; they are harmful to tropical fish. Due to these disadvantages, chloroamines were not used by many utilities until the THM discovery in 1974. After that, a large number of utilities changed from chlorination to chloramination, by simply adding ammonia ahead of chlorine feed to form chloramines.

Chlorine Dioxide Application

Chlorine dioxide, a yellow to red gas, is a very strong disinfectant produced by reacting sodium chlorite with chlorine or an acid (see Figure 10-3).

$$2NaClO_2 + Cl_2 \leftrightarrows 2NaCl + 2ClO_2\uparrow$$

$$5NaClO_2 + 4HCl \leftrightarrows 5NaCl + 4ClO_2\uparrow + 2H_2O$$

Both processes use a similar generator formed of a cylindrical reactor of polyvinyl chloride or Pyrex glass. The percentage yield/efficiency of the generator should be at least 95 percent. The efficiency is pH dependent; it is optimum at a pH of 3 to 5.

Electrolysis of sodium chlorite ($NaCLO_2$), another method, is still at the plant-scale testing stage; it is quite promising due to 100 percent efficiency. There is reduction of Na^+ to Na at the cathode and oxidation of ClO^-_2 to ClO_2 at the anode.

$$\underset{\text{Cathode}}{NaClO_2} \rightarrow \underset{}{Na} + \underset{\text{Anode}}{ClO_2}$$

The Niagara Falls, New York, water utility was the first to use chlorine dioxide in the United States in 1944 for taste and odor control. Since 1974, more than 400 water utilities in North America started using chlorine dioxide

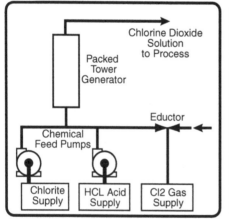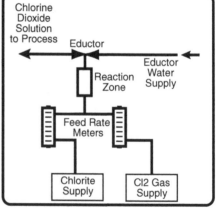

Figure 10-3   Chlorine Dioxide Generation

for predisinfection due to the THM formation problem with chlorine and CT requirements.

Chlorine dioxide is a strong disinfectant. Unlike chlorine, its effectiveness is not affected by ammonia and pH, and it does not produce THMs. Its effectiveness against *Giardia* and *Cryptosporidium* is reported as better than chlorine. It is very effective when followed by chlorine or chloramines. This sequential disinfection has a synergistic effect. A 1.5 mg/L chlorine dioxide dose alone gives only 90 percent reduction of these pathogens, whereas this dose followed by 1.6 mg/L chlorine or 2.8 mg/L chloramines gives about 99 percent reduction. However, chloramine and chlorine alone produce an insignificant effect. Apparently, chlorine dioxide weakens the pathogens and chlorine or chloramines destroy them.

Chlorine dioxide has disadvantages, however. It is an unstable compound that shortly reverts to chlorite; it is relatively expensive to generate; it is explosive at a concentration above 10 percent in the air; and it forms chlorites and chlorates. Being short lived, it is generated at the site and applied immediately. These disadvantages limit its application.

Byproducts, chlorites and chlorates, cause an anemic condition in some individuals, resulting in a limit on the chlorine dioxide dose. Activated carbon, sulfur dioxide, sulfite, and ferrous compounds have been tested to neutralize these byproducts. Out of these, the most practical and cost effective is the use of ferrous ions (Fe^{+2}). About 3 mg/L ferrous ion concentration reduces chlorite by 1 mg/L, a stoichiometric ratio. Ferrous chloride is the commonly used source of ferrous ions.

Ozonation

Ozonation is the use of ozone as a disinfectant. Ozone, a colorless gas with a peculiar pungent odor, is a triatomic form of oxygen, O_3. The smell is noticeable around electric machines. It is generated by either passing air or oxygen through a high-voltage arc between two electrodes, or by UV radiation. Commercially, it is generated by silent electric discharge or corona discharge by using a high-voltage (over 15,000 volts) electric current (see Figure 10-4). This process converts a small percentage of oxygen into ozone (1–3 percent from air, and 2–6 percent from the oxygen feed). Ozone is 20 times more soluble in water than oxygen, which helps in its dispersion in water. The maximum reported solubility of ozone in water is 40 mg/L.

$$3O_2 \leftrightharpoons 2O_3$$
$$\text{Oxygen} \quad \text{Ozone}$$

Ozone is an unstable gas; its molecules readily revert into oxygen molecules and nascent oxygen (single atoms). Its half-life at room temperature is only 15 to 20 minutes. Apparently, nascent oxygen causes disinfection.

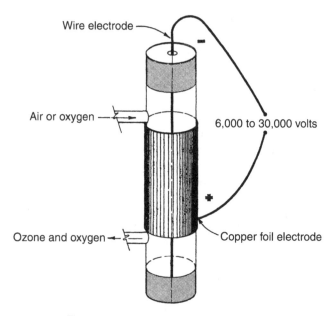

Figure 10-4 Electrical Discharge Ozonator

$$O_3 \rightarrow O_2 + O$$

Immediately after generation, ozone is dispersed through the water. Ozone kills microorganisms quickly. Their cells are ruptured (lysis). Ozone is the strongest known chemical water disinfectant. Ammonia and pH do not impair its effectiveness.

Ozone was first used for disinfection of the water in 1893 at the Oudshoorn water plant, Netherlands. It is more accepted in France, Switzerland, and Germany (more than 1,000 installations) because Europeans believe in treating their water to the pure, unpolluted, odorless, groundwater state. The first major installation of ozonation was in 1905 at Nice, France, followed by Paris in 1906. In the United States, it was used in 1906 for controlling tastes and odors in the Jerome Park Reservoir in New York City. After that, the first major installation was at Whiting, Indiana, in 1939, followed by Philadelphia in 1949. There is an increasing interest in ozonation due to stricter disinfection requirements and disinfection byproduct problems.

Ozonation treatment of water can be divided into three parts: preparation of feed gas, production, and contaction (see Figure 10-5).

- *Preparation* of feed gas is mainly filtration of air to remove dust particles and moisture. Moisture is corrosive to the system because it produces nitric acid in the generator by combining oxygen and nitrogen. Pure oxygen feed is more efficient because it does not have these problems.

Ozonation Process

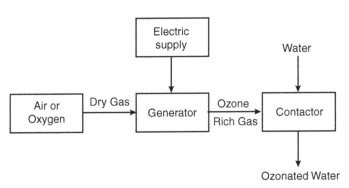

Figure 10-5 Ozonation System

- *Production* of ozone is the generation of ozone by passing the feed gas at a low pressure between the two electrodes separated by a dielectric and a gap, across which an alternating potential of about 15,000 volts is maintained for ozone generation. A large amount of heat is released in this process, which is dissipated into the cooling water.
- *Contaction* is contacting the microbes by the bubbling of an ozone and oxygen/air mixture through the water by dispersing it mostly by diffusers at the bottom of a contact chamber for the maximum ozone transfer.

Advantages of Ozonation. Ozone is an excellent disinfectant that also controls tastes and odors, color, algae, slime growth, trihalomethanes precursors, cyanides, sulfides, sulfites, iron, manganese, and turbidity. The required CT value of ozone is very low. Ozone is the strongest disinfectant, and monochloramine is the weakest. Table 10-1 shows the relative strength of disinfectants.

Table 10-1   Relative Strength of Disinfectants

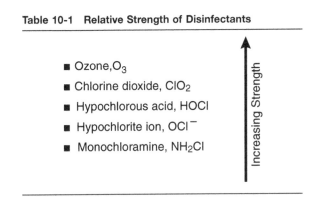

Disadvantages of Ozonation. There are some disadvantages of ozonation, such as the high cost of production, short-lived with almost no residual effect, on-site production, and difficulty in adjusting to variations in treatment load or demand. Furthermore, it causes fragmentation of large organic molecules, which encourage the bacterial growth in the distribution system; and it forms bromate, a THM precursor. Due to these shortcomings, ozonation is less common than chlorination. Generally, ozonation is followed by chlorination or chloramination for the residual effect.

Due to high capital cost and lack of residual effect, various combinations of ozone and other disinfectants, such as peroxone and soozone, have been tried. *Peroxone* is the use of hydrogen peroxide and ozone together. This combination accelerates the oxidation of some organics by two to six times when compared to ozone by itself. Hydrogen peroxide and ozone in a combination of 2 mg/L and 4 mg/L, respectively, are very effective. Peroxone treatment is almost half as expensive as ozonation alone. Peroxone treatment followed by 1.5 mg/L of chloramines has been proven by some utilities as an effective disinfection treatment. *Soozone,* a combination of ultrasonic waves and ozone, has been tried for some water with organic matter, where ultrasonic waves break organic particles, and ozone oxidizes them.

Ultraviolet Light Treatment

This approach disinfects water by ultraviolet (UV) light radiation. UV rays affect nucleic acids and enzymes of microorganisms and inactivate them. Unlike other disinfectants, UV treatment as a physical treatment does not produce harmful disinfection byproducts. In 1910, Marseille, France, used UV disinfection of drinking water. Currently, due to more stringent requirements, there is an increasing interest in disinfecting water with UV radiation. The most effective wavelength of UV light is 265 nm (nanometer = 10^{-6} mm). However, the most commonly available wavelength is 254 nm, which is produced by a mercury vapor lamp.Generally, low-pressure mercury vapor lamps are used for less than 1 mgd plants. Due to more interest in UV disinfection, low-pressure high output and medium-pressure lamps are developed for larger operations. These lamps are more expensive, but they are more effective and durable. For further improvement, pulse-wave-emitter mercury vapor lamps have been developed.

Based on the type of microbes and quality of the water, UV effective dose differs from system to system. Commonly, the drinking water is disinfected with 38 to 40 mj/cm^2 (milli joules/square centimeter) dose. A *joule* is a unit of energy that is equal to 0.239 calories. A 40 mj/cm^2 dose is reported to achieve 2 log, 99 percent, inactivation of *Cryptosporidium,* whereas 140 mj/cm^2 dose is needed for 99 percent reduction of enteric viruses. The higher the dose, the better is the disinfection, and more expensive is the treatment.

Factors Affecting UV Treatment

- *Turbidity.* Water should be as clear as possible because turbidity shields microorganisms from radiation. The higher the turbidity, the less is the transmittance, and less effective is the UV treatment.
- *Total dissolved solids.* Solids deposit on the lamp and foul it; therefore, the less the dissolved solids, the better is the treatment.
- *Dissolved organic matter.* Dissolved organic matter absorbs the UV light and shields microbes from radiation. The less the dissolved organic matter, the better is the treatment.
- *Depth.* The shallower the water, the better is the penetration of UV light, and more effective is the treatment. Therefore, for proper UV disinfection, water should be clear, colorless, shallow (3–5 inches deep), and stable to allow effective penetration of rays.

Considering all these factors, the best point for UV application is between filters and the clear well. These requirements, no residual effect, and high cost are some of the problems that need to be considered before changing to UV treatment. UV treatment is still quite expensive for most utilities.

CT AND CT RATIO CALCULATIONS

C stands for the concentration of a disinfectant as mg/L, which is the lowest residual value of the disinfectant at the highest flow at the effluent site. T stands for the contact time in minutes, which is the detention time of a pipe or a basin at the highest flow during 24 hours. CT value is $C \times T$ as mg/L-minutes. Thus, CT value takes into consideration the lowest concentration of the disinfectant and its shortest contact time. If concentration is high, then the lower contact time is required, and vice versa. CT calculation needs the disinfectant residual at the effluent end of each pipe and basin and their detention times at the highest flow to satisfy CT requirements.

The calculated CT for each part of the disinfection train divided by the required CT (from the CT table provided by the EPA, appendix B) for each disinfectant is called *available CT*. For compliance, the aggregate of all available CTs needs to be at least 1. The EPA and each state has given 2.5 log removal credit for *Giardia* and 2 log removal credit for enteric viruses to most of the utilities with conventional treatment plants. Therefore, those utilities need only 0.5 log (68 percent) removal of *Giardia* and 2 log (99 percent) removal of enteric viruses to satisfy the SDWA requirements. The higher the required log removal, the higher is the required CT value.

To eliminate the possibility of short circuits and thus a shorter contact time in the tanks and pipes, the EPA has provided a T_{10} multiplier factor table for each line or basin depending on its mixing characteristics (see Table 10-2).

Table 10-2 T_{10} Multipliers

Baffling Condition	Baffling Factor	Baffling Description
Unbaffled (mixed flow)	0.1	None, agitated basin, very low length-to-width ratio, high inlet and outlet flow velocities
Poor	0.3	Single or multiple unbaffled inlets and outlets, no intra-basin baffles
Average	0.5	Baffled inlet or outlet with some intra-basin baffles
Superior	0.7	Perforated inlet baffle, serpentine or perforated intrabasin baffles, outlet weir, or perforated launders (most filters are in this category)
Perfect (plug flow)	1.0	Very high length-to-width ratio (pipeline flow), perforated inlet, outlet, and intra-basin baffles

Due to thorough mixing, this factor is 1 for long transmission pipes (lines). For basins with better mixing due to better baffling and proper inlets and outlets sites, the factor is 0.7, and for the basins with some mixing, the factor is 0.5. Therefore, T value used for CT calculation of a pipe or a tank is its hydraulic detention time multiplied by T_{10} factor from the table. It reduces the T value of some basins to just half of their hydraulic value. The idea is to consider the worst possible conditions to satisfy CT requirements to ensure an adequate disinfection.

T_{10} stands for the time for 10 percent exiting of the water after entering the pipe or tank. Suppose 100 gallons of water enter a basin at a particular time. T_{10} is the time period when 10 gallons of that water are collected at the effluent end. T_{10} is determined by adding a chemical such as fluoride (tracer) at a specific time into the influent of a basin, and it is tested at short intervals at the effluent end. The time period it takes for 10 percent of the fluoride dose to reach the discharge end of the basin is marked. T_{10} factor for the basin is this time period divided by the hydraulic (actual) detention time of the basin.

For example, assume we use chlorine as the principal disinfectant to treat 30 mgd maximum flow at 20°C and 7.5 pH of water. The transmission line capacity (volume) is 0.5 million gal. The lowest free-residual chlorine at the effluent end of this line is 1 mg/L. After filtration, chloramines are added into the moderately baffled 1 million gallon reservoir, and the residual going to the distribution system is 2.5 mg/L. Calculate the CT value for the transmission line and the reservoir and their total available CT value. This plant needs only 0.5 log *Giardia* removal and 2 log removal of enteric viruses:

$$\text{CT calculated} = C \times T \times T_{10}$$

where:

C = the lowest disinfectant residual at the highest flow, as mg/L
T = detention time in minutes
T_{10} = multiplier

For transmission line

$C = 1$ mg/L

$T = ((\text{Volume}/\text{Flow}/\text{day}) \times 1440 \text{ min}/\text{day}) \times T_{10}$

$= ((0.5 \text{ MG}/30 \text{ mgd}) \times 1440 \text{ min}/\text{day}) \times 1 = 24 \text{ min}$

Calculated CT for transmission line = 1 mg/L × 24 min = 24.

Required CT from the table (appendix B) for 20°C and 7.5 pH for 1 mg/L of free residual chlorine is 11 for 0.5 log removal of *Giardia* and 1 for 2 log removal of enteric viruses. We are supposed to use 11, the higher of the two values, to satisfy the CT requirement.

Available CT for transmission line = Calculated CT/Required CT

Available CT = 24/11 = 2.18

Calculated CT for the reservoir

$C = 2.5$ mg/L

$T = (1 \text{ MG}/30 \text{ mgd}) \times 1440 \text{ min}/\text{day} \times 0.5 = 24 \text{ min}$

Calculated CT for the reservoir = 2.5 mg/L × 24 minutes = 60

Required CT for chloramine from the table is 185 at 20 C for 0.5 log *Giardia* and 321 for 2 log removal of viruses. The greater of the two values is 321.

Available CT ratio of basin = 60/321 = 0.19

Total available CT for transmission line and reservoir = 2.18 + 0.19 = 2.37

This plant is in compliance just by predisinfection. Most of the plants satisfy the CT requirement by proper predisinfection.

Note that the CT values in table are the lowest for ozone and highest for chloramines. The stronger the disinfectant, the lower is the CT value, and vice versa.

INDICATORS OF PROPER DISINFECTION

Turbidity

Turbidity below 0.3 NTU indicates very clear water and possibly the absence of any waterborne pathogens. The less the turbidity, the less the number of particles, and less is the possibility of the presence of pathogens.

Log Removal of Pathogens

The SDWA requires 3 log removal of *Giardia* and 4 log removal of enteric viruses for the compliance. *Log removal* is an engineering expression for percent removal in the logarithm (log) form. One log removal means 90 percent removal; 2 log removal means 99 percent removal; 3 log removal means 99.9 percent; 4 log removal means 99.99 percent removal, and so on. The following equation explains this conversion:

$$\% \text{ removal} = 100 - (100 \times \text{Log removal})$$

where:

Log = Negative power of base 10, e.g., 1 log is 10^{-1} or $1/10$ or 0.1, and 2 log is 10^{-2} or $1/100$ or .01

Example: Suppose we need to remove 4 log of enteric viruses.

$$\% \text{ removal} = 100 - (100 \times 10^{-4} \text{ or } .0001)$$
$$= 100 - 0.01 = 99.99$$

The higher the log removal, the less is the number of remaining pathogens in water, and better is the disinfection.

Coliform Bacteria

The absence of coliform bacteria means a possible absence of waterborne pathogens. They are required to be absent in 95 percent of all samples/month.

Heterotrophic Plate Count

The lower the count, the better is the disinfection. As a secondary standard, there should not be more than 500 bacteria/mL of the treated water.

CT Value

The higher the available CT value, the better is the disinfection.

Table 10-3 Disinfection Problems and Their Solutions

Problems	Possible Causes	Possible Solutions
Unable to feed sufficient chlorine.	Restricted feed line.	Check any obstruction in feed line and correct it.
	Chlorine tank close to being empty.	Change the tank.
	Chlorine leak. It is indicated by chlorine odors in the room.	Put on mask, apply ammonia fumes to detect the leak, and fix it.
	Frozen chlorine supply line could be due to too high a withdrawal rate from the tank.	Restrict daily withdrawal rate to 250 lbs./day as the maximum for 1,000 lb. cylinders.
Chlorine residual is low and there is no problem with the feed system.	Chlorine demand of the water has increased. Presence of more organic matter due to farmland runoffs, untreated sewage, and ammonia will cause higher demand.	Increase chlorine dose corresponding to the higher demand.
Chlorine residual is low and THMs are high.	Overfeeding the chlorine and reaching the breakpoint chlorination.	Check residual chlorine. If combined residual is decreasing and free residual is increasing, the free residual as HOCl is reacting with organic matter and producing THMs. Lower chlorine dose by about 0.5 mg/L or as required. It should lower the THMs. In the presence of THM precursors, always keep the free residual chlorine as low as possible. Ordinarily, it should be below 0.5 mg/L.
Chlorine residual is dropping when dose is being increased.	Overfeeding the chlorine and destroying the combined residual chlorine.	Lower the chlorine dose by 0.5 mg/L and check the residual. If residual comes up, then it was overfeeding. Continue cutting back the dose gradually until desired residual is achieved. It is important to understand breakpoint chlorination.
Ammonia odors are in chloraminated water.	Overfeeding ammonia.	Check the ammonia setting. If the setting is 100%; feed valve is wide open. Ammonia feed may be too high. If ammonia is too high stop ammonia feed and increase chlorine dose gradually to neutralize excess ammonia.

Table 10-3 (*Continued*)

Problems	Possible Causes	Possible Solutions
Chloramines are used for postdisinfection and there is low combined residual chlorine on the tap.	Chlorine tank is almost empty. It will be indicated by low chlorine residual and no effect on free-residual chlorine, ammonia, or pH.	Change the tank.
	Ammonia feed is low. It will be indicated by low ammonia content, pH slowly dropping, and free chlorine slowly increasing.	Check ammonia feed and adjust as required. Ratio of ammonia to chlorine should be 1:4.
Chloramines are used for postdisinfection and there is low combined residual chlorine on the tap. (continued)	Overfeeding the chlorine and reaching the breakpoint. This is indicated by decreasing combined residual chlorine, low ammonia, and increasing free-residual chlorine.	Reduce chlorine feed to 1:4 ammonia and chlorine ratio. Free residual will decrease and combined residual will increase.
pH of the tap is considerably lower than the filter effluent.	Excess chlorine when chloramines are used for postdisinfection.	Check chlorine residual. Free residual chlorine as hypochlorous acid lowers the pH. If free residual is close to total residual, then chlorine dose is too high. Lower it so the ammonia-to-chlorine ratio is 1:4 to keep it below the breakpoint. pH and combined residual chlorine will come up.
Chlorine dioxide generator efficiency is low.	pH of reactants in the generator is too high.	Check chlorine feed rate. Low chlorine feed will cause the high pH and lower efficiency. Chlorine dioxide generation is pH dependent. pH in the generator is around pH 4, or as recommended by the manufacturer. Increase chlorine feed rate until pH reaches the optimum. Further lowering of pH will have excess chlorine, which can cause THM formation. Above the optimum pH, there is an excess of residual chlorite that is also undesirable. pH of the generator must be maintained within a narrow recommended range.

Symptom	Possible Cause	Solution
There is low ozone residual from the generator.	There is an ozone leak. Ozone monitor should detect the leak.	Check for the leak around the seals and gaskets by using potassium iodide solution (a cloth soaked with the solution will work). Potassium iodide turns blue by reacting with ozone by producing iodine. Fix the leak. Tighten or repair the faulty connection.
There is a high ozone demand.	Low efficiency of the generator. Low efficiency will be caused by malfunction of the generator due to the impurity of oxygen feed by the high moisture contents of the air feed or burned generator tubes.	Check these parts of the system and correct them.
	Poor quality of the influent water. This condition is indicated by high turbidity, higher particle count, higher NOM content, and higher coliform or HPC count in the influent water to the ozone contact tank.	First, increase the ozone dose for adequate disinfection and then improve the treatment prior to ozonation. Prior to ozonation, water should be as clean as possible.

Minimum Required Disinfectant Residual in the Finished Water

In the distribution system, the minimum required chlorine residual is 0.2 mg/L, which ensures that water is adequately disinfected and its quality is protected in the distribution system.

Disinfection Problems and Possible Solutions

Refer to Table 10-3 for some common disinfection problems:

QUESTIONS

1. What are the two main sources of waterborne pathogens?

2. Name the hard-to-kill waterborne pathogens.

3. Which of these three are waterborne diseases?
   a. Malaria, flu, and hepatitis.
   b. Cholera, bacillary dysentery, and cryptosporidiosis.
   c. Amebic dysentery, dingo fever, and giardiasis.

4. Explain the term *log removal*. What is the log removal requirement for *Giardia* and enteric viruses for the SWTR?

5. State three reasons that make chlorine a popular disinfectant.

6. How much heavier is chlorine than air?

7. Any chemical that produces hypochlorous acid in water is considered chlorine. T or F

8. Explain breakpoint chlorination. What is the major advantage of breakpoint chlorination?

9. For 0.7 mg/L of ammonia dose, what is the corresponding chlorine dose?

10. Name the chemical that is used to detect chlorine leakage.

11. What are the two main objectives of postdisinfection?

12. What are the main advantages of chlorine dioxide treatment?

13. Name disinfection byproducts of each: chlorine, chloramine, chlorine dioxide, and ozone.

14. a. Ozone is a triatomic form of oxygen. T or F
    b. Ozone is the strongest known disinfectant. T or F

15. Which of these is the strongest disinfectant: chloramine, chlorine dioxide, or ozone?

16. Explain the term *CT*, and give its significance in the disinfection.

11

TASTE AND ODOR CONTROL

Besides sparkling clear, safe, and stable water, the public expects a good-tasting and odorless water as well. An average consumer judges the water quality from its looks, odor, and taste. Most water-quality complaints are about taste and odor, which are closely related. Mostly, a substance that produces an odor also produces a taste.

SOURCES OF TASTES AND ODORS

Decaying Organic Matter

Decaying of natural organic matter is one of the major causes of tastes and odors in surface water sources. Microbial decomposition of organic matter in sludge at the bottom of a reservoir, sedimentation basin, filter backwash well, and distribution lines causes odors during and after treatment. Generally, these odors are earthy, musty, moldy, or swampy. Most utilities have a good basin-cleaning and line-flushing program to control this problem.

Microbes

Algae and bacteria are important microbial groups that cause taste and odor problems.

Algae. Various groups and individual algae cause a variety of tastes and odors. After decaying dead organic matter, the algal group is the second most important cause of tastes and odors. Abundant growth of some species (algal blooms) causes earthy, musty, grassy, and fishy odors in the water. Blue-green algae, as a group, are probably the worst. They are common during summer until the early fall. Figure 11-1 shows the taste- and odor-causing algae.

Algae are controlled by using powdered activated carbon and algaecides, such as copper sulfate.

Bacteria. Bacteria are the major decomposing microbial group; therefore, wherever there is dead organic matter, there is also decomposition by bacteria, resulting in odors. Under anaerobic conditions, iron, nitrogen, and sulfur bacteria in the distribution lines produce rotten eggs, rusty, and fishy types of tastes and odors.

Actinomycetes, the branched bacteria, produce geosmin and methyl iso-borneol (MIB), which cause peculiar earthy-musty odors during spring and fall turnover. Odors of these compounds are noticeable at a very low concentration level (10 nanogram (ng)/L).

Metals

Iron, manganese, lead, copper, and zinc cause metallic odors in the water. These metals are either naturally present in the water or are contributed by corrosion in the distribution system. They are removed during treatment and controlled by corrosion control.

Disinfectants

Particularly, chlorine's taste is noticeable at 0.30 mg/L as free-residual chlorine, and 0.50 mg/L as chloramines. Postchlorination of a water treated with chlorine dioxide sometimes causes peculiar cat-urine-like odor in homes. Chlorine reacts with chlorites, a byproduct of chlorine dioxide, to reform chlorine dioxide in the distribution system. Chlorine dioxide reacts with some carpet material and causes these odors.

Sewage

Sewage causes septic odors. Sewage also fertilizes the receiving water bodies and stimulates algal growth.

CONTROL

Throughout the water treatment process, taste and odor removal has been one of the most serious concerns. Generally, taste and odor control starts at the source water and continues into pretreatment, coagulation, filtration, and post-disinfection.

Generally, tastes and odors are controlled by three methods: aeration, oxidation, and adsorption.

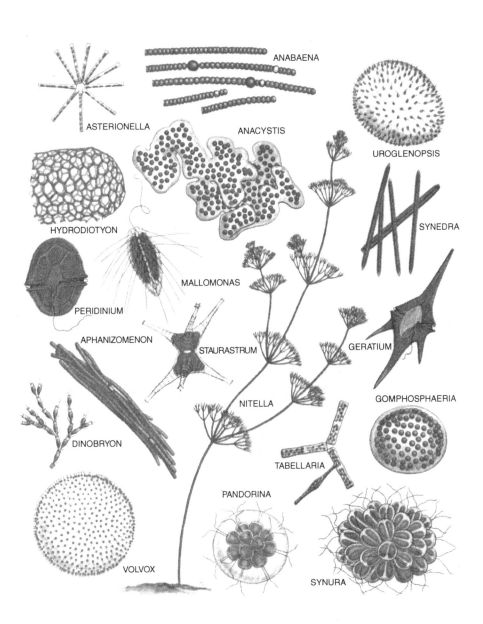

Figure 11-1 Taste- and Odor-Causing Algae
(*Source: Algae in Water Supplies,* U.S. Dept. of Health, Education, and Welfare, 1962.)

Aeration

Aeration is the passing of air through the water to degasify or oxidize odor-causing substances. It degasifies hydrogen sulfide and methane (marsh gas), and oxidizes iron and manganese.

Oxidation

Oxidation chemically neutralizes the taste- and odor-causing substances by using oxidants, such as potassium permanganate, chlorine, chlorine dioxide, and ozone.

Potassium permanganate (KMnO$_4$). This is a commonly used chemical for control of tastes and odors. It oxidizes taste- and odor-causing substances by producing nascent oxygen. Nascent oxygen (O) is the oxidant.

$$2KMnO_4 + H_2O \rightarrow 2KOH + 2MnO_2\downarrow \underset{\text{Nascent oxygen}}{+ 3O}$$

Potassium permanganate is available in its crystalline solid state as bags. It is fed by dry chemical feeders into a mixing tank and delivered as a solution to the feed point. Its solubility in water is about 5 percent (50,000 mg/ L). For best results, it is used into the raw water or into the effluent of presedimentation basin. Generally, 0.5 to 2.5 mg/L dose is quite effective. It is more effective at a high pH. An adequate dose is determined by jar testing. Apply only the optimum dose with proper detention time, because an overdose causes pink water condition.

Chlorine. In addition to disinfecting the water, chlorine oxidizes taste- and odor-causing chemicals, such as metals, hydrogen sulfide, and a number of organic compounds. Due to the THM problem, it is applied to presettled water after the removal of organic matter.

Chlorine Dioxide. Besides a strong disinfectant, chlorine dioxide is an effective taste- and odor-controlling agent. It controls musty and phenolic odors very effectively. It is applied to the effluent of the presedimentation basin. Because it does not produce trihalomethanes, it is a good substitute of chlorine for disinfection and for taste and odor control.

Ozone. Ozone, the strongest disinfectant, is also an excellent taste- and odor-controlling agent. Like chlorine dioxide, it is applied to the presettled water.

Adsorption

Adsorption is the acquisition or accumulation of substances, called *adsorbates,* on the surface of the adsorbing substance, called *adsorbant.* The most common adsorbant is activated carbon, and the taste- and odor-causing substances it removes are adsorbates. Activated carbon removes a large number of taste- and odor-causing substances. Activated carbon treatment and coagulation are two common means of adsorption in the water treatment.

Activated Carbon Adsorption. This is the main treatment for the removal of tastes and odors, volatile organic compounds, and synthetic (pesticides) organic chemicals. It has been known for a long time that charcoal filtration improves water's taste. Charcoal filters have been used since 1883 to make water taste better. Most of the point-of-use filters use activated carbon.

For water treatment, two forms of activated carbon—powdered activated carbon and granular activated carbon—are used. The choice depends on the source water quality and design of the plant.

Activated carbon is a wood product made by burning plant matter to form charcoal. Its formulation involves two processes, carbonation and activation. *Carbonation* is the burning of wood material at 550° to 700°C in the absence of air to form charcoal. *Activation* is the further burning of charcoal at 800° to 900°C in the presence of steam and carbon dioxide to produce pores and crevices in and on the surface of the particles. Pores and crevices increase the adsorbing surface tremendously and, thus, activate the adsorption. Activated charcoal is ground into a powder or into granules to form powdered activated carbon (PAC) and granular activated carbon (GAC), respectively. It is estimated that 1 pound of granular activated carbon has 2.5 to 9.0 million feet^2 area.

Powdered Activated Carbon Particle size of PAC ranges from 10 to 74 micrometers. The smaller the particle size, the quicker is the adsorption. Its density is 0.36 to 0.74 g/cm^3. It comes in a bulk load as powder. Powder is mixed in tanks with water, mostly at a rate of 1 pound/gallon of water to form slurry, which is kept mixed for dosing. Slurry is applied by means of volumetric metering and feeding system.

Slurry is added to the water at several points, starting from intake to the filter influent for the most effective treatment based on the quality of water and design of the plant. It is a *use and waste process;* therefore, it must be settled out in the sedimentation basins after adequate contact time. In addition to being a good absorbent, activated carbon is also a reducing agent because it reacts with disinfectants such as chlorine, chloramines, chlorine dioxide, and ozone. It neutralizes them and causes a higher disinfectant demand. Therefore, activated carbon and a disinfectant should never be used together.

For a proper contact time and no disinfectant interference, activated carbon is mostly applied into the influents of presedimentation basin and the final sedimentation basin. Other application points are rapid mix, flocculation basin, and influent of the primary sedimentation basin. It is applied to the influent of microfiltration and ultrafiltration processes. Besides disinfectants, its effectiveness is also reduced by the presence of an excessive amount of calcium carbonate in the water (e.g., unstabilized softened water). Calcium carbonate coats the carbon particles and blocks the adsorption sites. To prevent the calcium carbonate problem, carbon is applied after the recarbonation of the softened water.

A typical dose of PAC varies from 3 to 50 mg/L, depending on water quality and the required degree of treatment. An adequate dose is determined by jar testing. Generally, the jar test optimum dose is higher than the actual required dose. Thus, the operator should start with a lower-than-jar-test optimum dose to treat the water.

The advantages of PAC application are low capital cost, dose as required, flexibility of application point, always fully activated, and no required regeneration. It also works as a coagulation aid by providing particulate matter for flocculation. However, it produces extra sludge.

Granulated/Granular Activated Carbon. This method is used for a continuous treatment when there is a need for continuous removal of volatile and synthetic organic compounds in addition to occasional taste and odor control. It is used as multimedia carbon filters or monomedium contactors. It does filtration and adsorption. The reaction zone is 5 to 6 inches deep top carbon layer; therefore, a minimum of 12 inches in the bed is needed. For the maximum effect, two feet depth is recommended. Proper adsorption requires at least 3 to 4 minutes contact time.

The preferred effective particle size is 0.35 to 0.45 mm (although it is available as 0.55 to 0.75 mm and even 0.85 to 1.05 mm). As GAC neutralizes any disinfectant on the surface, there can be bacterial build-up inside the particles and in the filter bed. To prevent the bacterial growth, a single dose of 2 mg/L of free residual chlorine in the backwash water is applied. After the GAC is exhausted (meaning all adsorption sites are completely used up); it starts releasing adsorbed substances. The releasing is known as *desorption*. At this point the GAC needs to be regenerated. Regeneration is almost like its formulation. For regeneration, medium is taken off the filter, dried at 100°C, pyrolysed at 650° to 750°C, and reactivated at 870° to 980°C in the presence of steam.

GAC treatment is more expensive than PAC treatment due to regeneration and the initial cost to build contactors. Furthermore, it does not give the flexibility of application at different points. Due to these reasons, most plants use PAC rather than GAC.

Pesticide Removal

Activated carbon treatment is the only effective treatment to remove pesticides such as atrazine and other synthetic organic chemicals. According to the EPA, GAC is the best available technology to remove volatile and synthetic organics. PAC removes atrazine from 8 to 9 ppb (micrograms/L) in the source water to less than 3 ppb (MCL for atrazine) by applying 12 to 18 mg/L dose.

Besides all the other good qualities, water should taste good. Drinking a glass of water should be enjoyable rather than just a necessity.

Table 11-1 Tastes and Odors Problems and Their Solutions

Problems	Possible Causes	Possible Solutions
There is excessive scaling of carbon slurry tank and feed lines.	Dilution water is very hard. Calcium carbonate and carbon deposit together to form hard and compact deposits.	If possible, use soft water.
	pH of carbon slurry is very high.	Check with manufacturer to know if carbon is acid washed. If not, lower pH of slurry by adding dilute hydrochloric acid into slurry tank.
Carbon causes high disinfectant demand.	Carbon is reacting with disinfectant. Reaction will be indicated by less effectiveness of activated carbon.	Always feed activated carbon; settle it out; & then, apply disinfectant. Check feed points & change them as required.
Suddenly, there is an earthy-musty odor in the water.	If it is spring or fall and air temperature is close to 40°F, these odors are due to the turnover. Odors are caused by decomposition of dead organic matter by bacteria, actinomycetes, and fungi. These odors are best controlled with powered activated carbon.	Determine and apply the appropriate dose of carbon by jar testing. Mostly, these odors need a 5–10 mg/L powdered activated carbon dose in the final sedimentation basin after recarbonation.
	Heavy rain upstream and runoffs.	Run a jar test; determine optimum dose of activated carbon, and apply the activated carbon.
Fishy odors are in the water.	May be due to algae or algal decomposition in the source water. *Peridenium*, a fire alga, is a serious offender. *Volvox*, a flagellated green alga, decomposes to cause these odors.	Check source water for the type and populations of algae by concentrating a water sample, preparing a slide, and studying under a microscope.
	Sludge is decomposing at the bottom of basins.	Smell the sludge coming out of basins; remove sludge until sludge smells normal. Apply required carbon dose.

139

Table 11-1 *(Continued)*

Problems	Possible Causes	Possible Solutions
Water has cucumber odors and metallic dry taste.	Algae like *Synura and Peridenium*, cause these odors, which are noticeable and offensive. These odors contribute to customer complaints if not controlled.	Check the type of algae in source water. Control at source, if possible. Run jar test; determine required carbon dose, and apply.
High atrazine is in the presettled water.	High atrazine level in the source water. This pesticide is used to control weeds, especially in wheat and sorghum fields. After heavy rain, it enters source water through runoffs from farm land.	Monitor atrazine level in source water. Determine adequate dose of activated carbon by jar testing and apply that dose.
	Low carbon slurry concentration.	Adjust the dose that corresponds to the slurry concentration. Monitor slurry concentration, especially, after getting a new load or after diluting it.
Customer water smells like cat urine.	Chlorite in water. It happens if chlorination follows chlorine dioxide treatment. Chlorine reacts with chlorites which are the byproduct of chlorine dioxide. Chlorine dioxide reacts with certain carpets to cause the problem.	Reduce chlorites either by lowering chlorine dioxide dose or by using a proper dose of ferrous.
	Ferrous feed system is not working.	Check the feeder setting and concentration of ferrous solution, and adjust the setting.

TASTE AND ODOR PROBLEMS AND POSSIBLE SOLUTIONS

Refer to Table 11-1 for some common taste and odor treatment problems and solutions.

QUESTIONS

1. State three causes of tastes and odors.

2. What causes the earthy-musty odors?

3. Most customer complaints are about taste and odor. T or F

4. What is the major difference between oxidation and adsorption?

5. Why is the use of chlorine limited in removing tastes and odors?

6. What are the major advantages of powdered activated carbon over granular activated carbon?

7. What water conditions require the use of GAC rather than PAC?

8. Differentiate between carbonation and activation. What are the appropriate temperatures for these processes?

9. How are pesticides and volatile organic and synthetic organic compounds effectively removed from the water?

12

FLUORIDATION

Fluoridation is the use of fluoride in the drinking water. Fluoride is an important component of bones and teeth. Fluoride deficiency causes weaker bones and tooth decay (dental cavities). Too much fluoride causes skeletal and dental fluorosis, resulting in brittle bones and mottled teeth, respectively. Therefore, just the right amount of fluoride supply is important for healthy bones and strong teeth. An effective daily dose of fluoride is 0.9 to 1.7 mg/L. A dose less than 0.7 mg/L does not do the job, and more than 4.0 mg/L can cause fluorosis. The U.S. EPA has set the maximum contaminant level for fluoride in the drinking water at 4 mg /L.

We get our fluoride supply from the drinking water and from treatments such as use of fluoridated toothpaste, fluoride solution, fluoride tablets, and fluoride drops.

FLUORIDE IN DRINKING WATER

Fluor spar (calcium fluoride (CaF_2)) is a common mineral present in all soils. It is carried by rainwater into drinking water sources. Some fluoride comes from sewage due to previously fluoridated drinking water. Fluoride is a very stable and persistent substance. Although naturally present in almost all source water, fluoride is mostly less than 1 mg/L. Thus, fluoridation of drinking water provides the remaining fluoride requirement.

Fluoridation of drinking water is not required by law; therefore, it is an extra service provided by the water utility to the community. Fluoridation of water has been a controversial issue due to health effects of higher amounts of fluoride and perhaps also due to the use of sodium fluoride as a pesticide.

People who have used sodium fluoride as a rat or insect poison object to the deliberate use of this substance in their water. Fluoridation of drinking water needs to be authorized by the city council or other local governing body.

At present, about 60 percent of the U.S. population is provided with an adequate amount (about 1 mg/L) of fluoride in its water supply. In the United States, fluoridation started in 1945 at Grand Rapids, Michigan. Use in other water utilities soon followed. Fluoridation has reduced dental cavities by 50 to 65 percent. For proper health protection, a person needs a continuous and adequate daily supply of fluoride; otherwise, the effectiveness ceases as the supply stops.

It is a sure, more convenient, easy, and economical way to have a continuous fluoride supply through drinking water than through other sources. About 50 percent of the fluoride supply is immediately absorbed by the enamel (outer hard) layer of teeth, and the rest is taken by the blood to different parts of the body to help the bones. It costs less than 25 cents/person/year to the water utility to fluoridate the water.

FLUORIDE APPLICATION TO THE WATER SUPPLY

Fluoride is available in 50-pound bags of crystalline sodium salts, such as sodium fluoride (NaF) and sodium silicofluoride/sodium fluosilicate (Na_2SiF_6) and in liquid drums as fluosilicic acid (H_2SiF_6). Sodium silicofluoride is the most common, economical, convenient, and safe fluoride source. It is applied through dry chemical feeders. The dose is continuously controlled, monitored, and recorded by the feed system with an alarm for an over- and underfeed. An accurately controlled dose is fed into a mixing and dissolving tank to make a homogeneous solution. Thus, an accurate fluoride solution dose is fed to the treated water—generally, the filter effluent. For further dose control, fluoride residual is monitored on line and by laboratory testing every three or four hours.

To determine the fluoride dose, background fluoride in the water is taken into consideration. Suppose we need 1.2 mg/L fluoride in the water and the background fluoride is 0.5 mg/L. Setting of the feeder needs to be 1.2 mg/L minus 0.5 mg/L or 0.7 mg/L as fluoride (not as fluoride compound). For example, sodium silicofluoride is used. It is only 60.7 percent fluoride. To feed 0.7 mg/L, we need to feed 0.7/0.607 mg/L or 1.15 mg/L of Na_2SiF_6. Ideally, fluoride residual in the treated water should be 1 ± 0.1 mg/L.

SAFETY ISSUES

Ingestion and inhalation of fluoride dust is poisonous. When swallowed, fluorides cause vomiting, stomach cramps, and diarrhea. Always avoid fluoride

Table 12-1 Fluoridation Problems and Their Solutions

Problems	Possible Causes	Possible Solutions
Low fluoride reading when the fluoride dose is correct.	Feeder is feeding improperly.	Check and correct the setting.
	Fluoride has high impurity.	Check the percent purity of the fluoride powder and correct the setting.
	Incomplete dissolving of fluoride powder. It will be indicated by settled out fluoride in the tank. It can be caused by improper mixing, low dilution water volume, and the lower water temperature.	Check all the factors and correct them. For lower temperature, allow longer detention time.
	Low background fluoride level.	Check the background level daily and dose accordingly.
	Analysis is incorrect.	Use an electrode or ion chromatography fluoride analysis method. Alum and iron interfere in the colormetric method.
There is build-up of fluoride in feed lines.	Water is very hard. Hard water will produce CaF_2, which forms scale in the lines.	Use softened water for making a fluoride solution and clean lines. Maintain clean lines.
Feed lines are corroded.	Metal pipes are corroding.	Fluoride solution is very corrosive; therefore, use corrosion-resistant metals or plastic pipes as recommended by the manufacturer.

dust. While handling fluoride, the operator should always use protective gear, such as mask and coveralls, especially when filling the feeder bin or cleaning the floor of the fluoride room.

DEFLUORIDATION

It is the removal of an excessive amount of fluoride from the water. Fluoride minerals can cause higher amounts of fluoride in the groundwater. Excessive amounts of fluorides are also caused by volcanic action and ocean water. Many groundwaters in the United States contain excessive levels of fluoride (e.g., coastal region of South Carolina's Black Creek aquifer has 2 to 6 mg/ L of fluoride). Another region with high fluoride water content is the vertical band of states from North Dakota to Texas.

There are two common methods of fluoride removal from the water: reverse osmosis membrane filtration and ion exchange treatment. Reverse osmosis membrane filteration has been used by Mount Pleasant, South Carolina, to reduce the fluoride level in the groundwater from 6 mg/L to 1 mg/L. Other utilities in the coastal region also use reverse osmosis. The ion exchange method is practical only for individual water supplies and industrial applications. Activated alumina (Al_2O_3) beads are very effective for ion exchange.

FLUORIDATION PROBLEMS AND POSSIBLE SOLUTIONS

Refer to Table 12-1 for some common fluoridation problems and their solutions.

QUESTIONS

1. Why is fluoride important for our health?

2. Why is the water supply the best source of community fluoride supply?

3. Why is fluoridation a controversial issue in some communities?

4. What is the effective dose of fluoride in water, and what is its optimum dose?

5. What is the common source of our fluoride supply?

6. State the health effects of too much and too little fluoride in the water.

7. Why do we need a continuous and effective dose of fluoride?

8. What causes an excessive amount of fluoride in groundwaters?

9. How is excess fluoride removed from the water?

13

WATER TRANSMISSION

Water transmission is the delivery of water from the source to the treatment plant and from the treatment plant to consumers through the distribution system. A distribution system is a network of water lines (pipes) that deliver treated water to consumers. A transmission system consists of the following major components:

- Pipes (transmission lines) for delivering water
- Pumps for pumping and maintaining pressure
- Valves to control water flow
- Reservoirs and elevated tanks for water storage
- Meters to measure the quantity of water supply
- Fire hydrants to provide sufficient pressurized water for firefighting

PIPES

Pipes deliver water to different places. Commonly, pipes are formed of cast iron, reinforced concrete, plastic, or copper. Cast iron pipes are ductile or gray cast iron type. A ductile iron pipe differs from a gray cast iron pipe in that it is malleable (can be hammered into a sheet); therefore, it can withstand more stress. Cast iron pipes are corrosion resistant, strong, and durable. For further corrosion control, they are lined with cement and tar. They are, mostly, larger than 2 inches in diameter, used for such things as lines to and from pumps. Reinforced concrete pipes are very large. Plastic pipes are light, non-corrosive, and available in sizes 1/4 to 8 inches. Copper lines, due to their flexibility, are used for service connections and for plumbing.

Pipe size depends on population and future development of a community. Mostly it is possible to use 6- to 8-inch pipe to supply a residential area. Domestic service lines are generally 3/4- to 1-inch plastic or copper; commercial lines are 2 inches.

PUMPS

A pump is a device for transportation of a liquid from one location to another, for lifting the liquid to a higher elevation, or for increasing the liquid pressure. Water is taken from the source to the treatment plant and transported to the customers, under pressure, by the pumps. Good understanding of pumps and pumping is important for an adequate pressurized water supply for domestic use, for commercial use, and for firefighting.

Pumping Conditions

Pumps operate under various conditions, such as static head, dynamic head, pumping head, suction lift, suction head, and friction loss.

Total static head is the vertical distance, as feet, from the free level of the source water to the free surface of discharged water.

Total dynamic head is the total static head plus friction head. Dynamic term is used when water is moving against the friction head. Simply add the friction head to the static head to determine the dynamic head.

Dynamic head = Static head + Friction head

Suction lift lifts the water from its source to the center of the pump (see Figure 13-1). It exists when the source water is below the center line of the

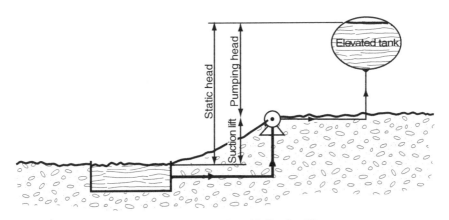

Figure 13-1 Pumping with Suction Lift

pump; thus, the pump has to lift the water before pumping it to the discharge point. Suction lift is the height between the free surface of the source water and the center line of the pump. For total static head, it is added to the pumping head, which is the distance between the center line of the pump and free surface of the discharged water. Standard atmospheric pressure is 14.7 pounds per square inch (psi), or 34 feet head of water. Theoretically, a pump can lift water to a height of 34 feet due to atmospheric pressure. Due to inefficiencies, friction losses, and bends in pipes, however, pumps have only 15 to 22 feet of practical suction lift.

Suction head is present when the free surface of source water is above the center line of the pump (see Figure 13-2). It is the height between the center line of the pump and the free surface of the source water. In this case, the total static head is the pumping head minus the suction head.

Total static head = Pumping head − Suction head

Keep in mind that the total static head is always the distance between surface of source water and the surface of the discharged water.

There are three main classes of pumps: centrifugal, displacement, and air-lift. The following sections discuss each class.

Centrifugal Pumps

Centrifugal pumps are the most commonly used pumps in the water utility; thus, they are discussed in detail and other types only briefly. A centrifugal pump has an impeller that rotates in a casing, sucks water, and slings it out by centrifugal force. These pumps are mostly used for lifting water from lower to a higher elevation, boosting pressure in the distribution system, and for transportation of water to and from the treatment plant.

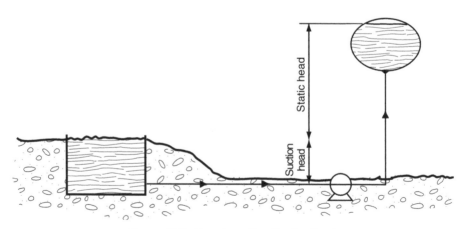

Figure 13-2 Pumping with a Suction Head

Figure 13-3 shows the parts of a centrifugal pump:

- Impeller
- Shaft
- Casing
- Bearings and packing
- Motor

Impeller is the rotating disc with vanes of various shapes for different types of flow, such as axial, radial, or mixed, depending on the discharge flow type. Rotation of the impeller creates the centrifugal force, which creates a vacuum at the eye of the impeller. Water, then, is sucked from the source due to the atmospheric pressure and forced out to the discharge point (see Figure 13-4). Impellers, with open, semienclosed, and enclosed vanes, cause a *radial flow,* which means water is discharged at a right angle to the suction. An impeller with vanes like a propeller causes an *axial flow,* which sucks water and forces it out in a straight line. *Mixed flow,* or *Francis flow,* is caused by a conelike impeller with spiral vanes discharging water at a 45° angle from the suction; it is a combination of radial and axial flows. Impeller types are illustrated in Figure 13-5.

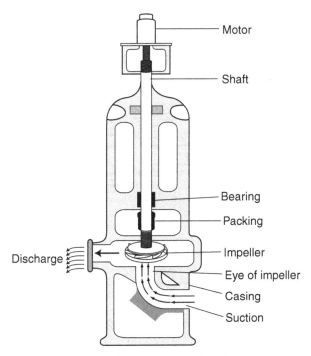

Figure 13-3 Vertical Section of a Centrifugal Pump

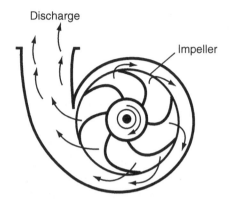

Figure 13-4 Impeller Rotation and Discharge

Shaft is a steel, stainless steel, or bronze rod to which an impeller is attached at one end and a motor, the prime driving force, is attached at the other end. The shaft transmits power from the motor to the impeller. It needs to be as straight as possible to reduce the friction losses and vibrations, which cause the higher motor loading; the shaft is kept in line by ball or roller bearings.

Casing is the housing for the impeller with the suction and discharge openings. Water enters at the suction opening and leaves at the discharge opening. The impeller housing is sealed off from the atmosphere, with a *packing* or a *seal,* to prevent air from getting into the water and excessive water leakage. The packing is formed of braided or woven cotton impregnated with graphite and metallic plastic. Some manufacturers use mechanical seals instead.

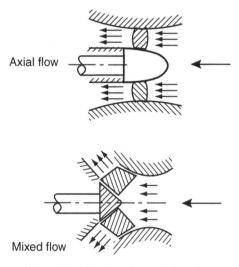

Figure 13-5 Various Types of Impellers

The *motor* is the prime mover of the shaft. It is connected to the shaft with a flexible coupling to reduce excessive wearing of shaft and bearings.

Pump capacity is the flow rate or the amount of water delivered per time unit, such as gallon per minute (gpm). A centrifugal pump's capacity is mainly controlled by two factors, revolving speed of the impeller and its diameter. Basic principles can be summarized into three rules:

1. Q, the flow rate (capacity), varies directly with the speed (revolutions per minute, RPMs) of the impeller or the impeller diameter.

$$Q_1/Q = S_1/S, \text{ so } Q_1 = Q \times (S_1/S)$$

where:

Q = flow rate for speed S
Q_1 = flow rate for speed S_1

2. H, the head (pressure), is directly proportional to the square of the change in speed or the diameter of the impeller.

$$H_1/H = (S_1/S)^2, \text{ so } H_1 = H \times (S_1/S)^2$$

where:

H = head for speed S
H_1 = head for speed S_1

3. Break horsepower (BHP), the power input, is directly proportional to the cube of the change in speed or the diameter of the impeller.

$$P_1/P = (S_1/S)^3$$

$$P_1 = P \times (S_1/S)^3$$

where:

P = power input for speed S
P_1 = power input for speed S_1

Example. A pump is rated 700 gpm for 15 ft. head. It requires 5 HP at 1,500 RPM. What will the flow rate, head, and power consumption be when speed is changed to 2,000 RPM? By applying the equations from the basic principles, we arrive at these answers:

Q_1 = 700 gpm $\times$ (2000 RPM/1500 RPM) = 700 gpm $\times$ 4/3 = 933 gpm

H_1 = 15 ft. $\times$ $(4/3)^2$ = 15 ft. $\times$ 1.8 = 27 ft.

P_1 = 5 HP $\times$ $(4/3)^3$ = 5 HP $\times$ (64/27) = 11.9 HP

Similarly, an alteration in the diameter of the impeller changes the flow, head, and brake horsepower. To calculate the changes due to the diameter, substitute diameter for speed in the previous equations. That is:

$$Q_1/Q = D_1/D, \text{ so } Q_1 = Q \times (D_1/D)$$

where:

Q = flow rate for diameter D
Q_1 = flow rate for diameter D_1

$$H_1/H = (D_1/D)^2, \text{ so } H_1 = H \times (D_1/D)^2$$

where:

H = head for diameter D
H_1 = head for diameter D_1

$$P_1/P = (D_1/D)^3, \text{ so } P_1 = P \times (D_1/D)^3$$

where:

P = power input for diameter D
P_1 = power input for diameter D_1

Example. If a pump at a certain speed is rated 300 gpm with 120 ft. head and requires 20 HP with a 6 inch impeller diameter. What will be the flow rate, head, and the HP if the diameter of the impeller is 7 inches?

Q_1 = 300 gpm $\times$ (7 inch/6 inch) = 300 gpm $\times$ 7/6 = 350 gpm
H_1 = 120 ft. $\times (7/6)^2$ = 12 ft. $\times$ (49/36) = 163.3 ft.
P_1 = 20 HP $\times (7/6)^3$ = 20 HP $\times$ (343/216) = 31.76 HP

Manufacturers provide a performance curve for each pump that describes its performance characteristics by giving the relationship of capacity, head, efficiency, and brake horsepower (see Figure 13-6). After reaching the highest head, the water simply churns in the discharge pipe, without any discharge. This highest head is called the *shutoff head*.

Classification of Centrifugal Pumps. Classification of centrifugal pumps is based on the services they perform, the type of flow, and the type of casing.

• *Low-service pumps* lift water from the source to the treatment plant and from clear well to the filter backwash system.

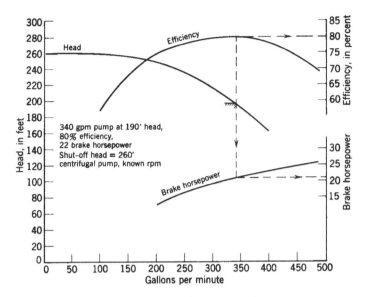

Figure 13-6 A Centrifugal Pump Curve

(*Source:* Salvato et al., *Environmental Engineering, Fifth Edition.* Copyright © 2003 by John Wiley & Sons, Inc. Reprinted by permission of John Wiley & Sons, Inc.)

- *Well pumps* lift water from the well and discharge it to the treatment plant. They are multistage centrifugal pumps for high head.
- *High-service pumps* transport water under high pressure from the treatment plant to the distribution system and to the elevated tanks. For very high elevations, two or more similar pumps are used in a series with the dischage of one entering the suction of the next. Suppose one pump delivers 200 gpm at 25 ft. head. For delivering 200 gpm to 100 ft. head, four pumps or stages (100/25) would be needed.
- *Booster pumps* boost the pressure in the distribution.
- *Backwash pumps* are used for backwashing the filters.
- *Sampling pumps* send samples to auto analyzers and laboratory.
- *Axial flow pumps* have suction and discharge in the same direction. They have a propeller-like impeller.
- *Radial flow pumps* discharge water at a right angle to the suction.
- *Mixed-flow pumps* have dischage at 45° to the suction
- *Volute pumps* have the discharge pipe diameter gradually increasing, which decreases the velocity and increases the pressure. They have spiral-shaped guiding vanes on the interior of the casing, called the *volute*. They are used for high lift, booster, or service pumps.
- *Turbine pumps* are diffused flow pumps. The impeller is in the center of the circular casing that has fixed vanes to reduce the velocity and increase the pressure.

- *Screw centrifugal pumps* have a screw-shaped impeller that combines the characteristics of the screw pump and a centrifugal pump. They are very efficient for high heads and water with particulate matter like sand.

Advantages of Centrifugal Pumps

- They are simple, light, and compact.
- They are vertical or horizontal.
- They are quiet.
- They produce uniform discharge.
- They require low maintenance.
- They don't have chambers with valves.
- They are not damaged when operated with discharge valve closed.

Disadvantages of Centrifugal Pumps

- They have low efficiency due to friction loss and slippage of water.
- They are not self primed.
- They have low suction lift, only 15 feet.
- They have a problem of air leaks into the casing causing *cavitation,* which is the formation of air pockets due to the fusion of small bubbles. Cavitation causes corrosion.

Displacement Pumps

Displacement pumps have a mechanism such as a piston, plunger, diaphragm, gears, cams, or a screw to force the liquid through the pump. They are used, mainly, for feeding chemical solutions and can be divided into two classes:

1. Reciprocating
2. Rotary

Reciprocating pumps have a mechanism to fill a chamber with liquid and to force it out by using a plunger or a piston. They create lift and pressure by displacing liquid. The volume or capacity of the pumped liquid is constant regardless of head. Chemical feed pumps may have two pistons or double-acting plungers that discharge liquid on both the forward and the backward strokes. Plunger pumps are useful for small quantities and high heads. They may have a single- or double-acting cylinder with one chamber filling and other discharging with the same stroke of the piston (see Figure 13-7). Capacity varies by speed (number of strokes/min) and stroke size (percentage of stroke). These pumps are used for pumping chemical solutions such as liquid chlorine, liquid alum, polymers, and potassium permanganate.

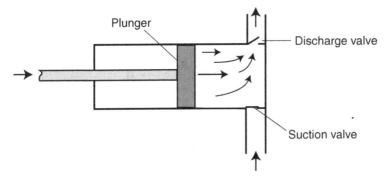

Figure 13-7   Plunger Pump

Advantages of Reciprocating Pumps

- They have high suction lift (up to 22 ft.).
- They are self-priming.
- Capacity is not affected by the head.

Disadvantages of Reciprocating Pumps

- They are mostly heavy.
- They are bulky with slow speed (50 to 150 strokes per minute).
- They produce pulsating flow.
- Cost is a concern, as they are expensive.
- A closed discharge valve can cause serious damage to the pump.

Rotary pumps are quite simple in design. They have two cams or gears; one is connected to the shaft and is called the *driving gear* that drives the other gear, the *idler gear.* A close-fitted casing surrounds the gears and has the suction and discharge ports (see Figure 13-8). The liquid fills the space around the gears. When the gears rotate, they mesh and squeeze the liquid out to the discharge port. When teeth separate, they create a partial vacuum and suck more liquid. Thus, the liquid is sucked in and squeezed out alternatively by each rotation. Like the reciprocating pumps, these pumps are self priming, with 22 feet suction lift, and a relatively high speed up to 1,750 RPMs.

Screw pump is a rotary pump that uses a screw-like pumping structure around a shaft instead of gears or cams. The screw rotates in a water-containing chamber and carries the water in a progressive cavity to the discharge point.

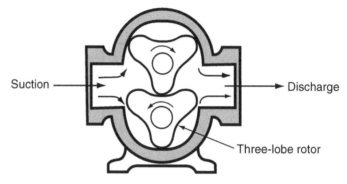

Figure 13-8   Rotary Pump

Air Lift Pumps

Air lift pumps use compressed air to lift abrasive sandy water. Compressed air is mixed with the water to decrease water density and increase water volume. Thus, the water rises to the discharge port, gets pumped out, and then more water is sucked in (see Figure 13-9).

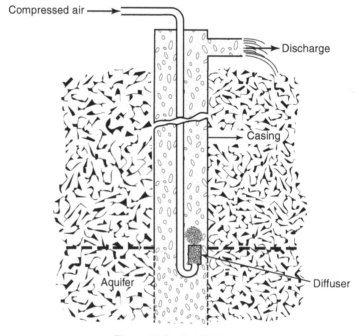

Figure 13-9   Air Lift Pump

VALVES

Valves are devices that control the direction and amount of flow of a liquid. They are used in water lines, tanks, and feed lines. There are four types of valves, based on their application.

- Flow control valves
- Check valves
- Pressure control valves
- Pressure and air relief valves

Flow Control Valves

Flow control valves consist of a gate valve, butterfly valve, and a needle valve. *Gate valves* are used to control flow from one part of the distribution system to another. They are used to isolate a section of line for repair or to isolate a part of the distribution system. Like the name, a gate valve is a gatelike disc fitted tightly in a groove inside the valve housing to control the flow (see Figure 13-10). To open or close, the gate is operated up and down by a hand wheel or by a motor operation. When fully open, there is no resistance to the flow. For convenience, these valves (in the distribution lines) are located about 500 feet apart in a business area and 800 feet apart in a residential area.

A *butterfly valve* is a disclike plate that rotates around an axis or shaft to open or close the valve to any degree (see Figure 13-11). Unlike a gate valve, it offers some resistance to flowing water. Butterfly valves are used to control the flow rate.

A *needle valve* consists of a long conical structure that is fitted into a conical housing. The structure moves up and down, allowing a gradual opening and closing of the valve by controlling the space between the needle and the housing. Needle valves are used in some chemical feeders (e.g., ammonia feeder). A needle valve should never be fully opened because solution pressure can cause it to feed a higher dose than its maximum rating.

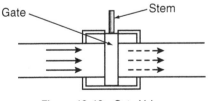

Figure 13-10 Gate Valve

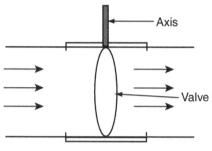

Figure 13-11 Butterfly Valve

Check Valves

Check valves are close-fitting platelike structures that are used to control the direction of the flow. They are swing type in horizontal pipes and lift type in vertical pipes. A lift-type check valve is called a *foot valve* when located at the bottom of a suction lift line of a pump to keep the pump primed (full of water). It is a disc hinged to the wall of a pipe and opens only in the direction of suction.

Pressure Control Valves

Pressure control valves are used to regulate the pressure between the high and low pressure points of a system. They are called *altitude valves* when installed in the inlet pipes of reservoirs, elevated tanks, and stand pipes to keep the water level within the operational levels. They stop the flow when water reaches the upper set level and start the flow when water level drops to the lower set level.

Pressure Relief and Air Relief Valves

Pressure relief and air relief valves are used for releasing excessive water pressure, such as due to water hammer and pressure due to air trapped in the lines, respectively.

Pumps and valves are the important parts of a water utility for a controlled transmission of water.

RESERVOIRS AND ELEVATED TANKS

Reservoirs and elevated tanks are used for storage of water and adequate pressure. Every water system keeps a sufficient amount of water for emergencies, such as breakdown of treatment or firefighting.

Reservoirs

Reservoirs provide storage for ground-level water. The reservoir at the treatment plant, which receives the filter effluent, is called a clear well. Other reservoirs are located at different points in the distribution system for an adequate supply of water. They are reinforced concrete or steel structures. Concrete reservoirs are built partially underground. The reservoir capacity is generally sufficient to meet the water supply for 4 to 6 hours. Advantages of reservoirs are that they provide a reserve supply of water, uniform pumping, proper detention time for disinfectant, and mixing of water from different sources. For proper protection, a reservoir should be partly above ground level, away from any sewer line (within 50 feet, the sewer line should be extra heavy cast iron), crack free, covered, and equipped with overflow capability, vent, drains, and manholes. It must be painted inside and outside to prevent corrosion. Vents and overflows are screened with fine mesh to prevent the entry of birds and insects. The drain of a reservoir should never be directly connected to the sewer lines.

Elevated tanks

Elevated reservoirs provide pressure due to height. The main advantage of an elevated tank is the availability of reserve water under pressure without pumping. Usually, water at a rate of 60 gallons/person/day must be elevated for fire protection. Minimum required capacity of an elevated tank is 50,000 gallons. Generally, an elevated tank is located close to high-consumption and high-value areas for better fire protection. An elevated tank should be properly maintained by painting inside and outside. It should also have cathodic protection and be covered.

They are connected to the distribution system and filled when water demand is low and emptied when demand is high. The emptying tank is called a *floating* tank.

METERS

Meters are the devices to measure the quantity of water used by the customer. Meters may be displacement, current, proportional, or compound in type.

Displacement Meters

Displacement meters are used for domestic customers. A displacement meter has a measuring chamber of a definite capacity. As the chamber fills and empties while the water flows through it, the meter records the quantity of that water measured by a nutating disc, oscillating piston, or a rotating gear. They are good for low flows.

Current Meters

In current meters, the velocity of the passing water causes a bladed wheel to rotate and to record the volume. The higher the velocity, the more is the volume, and vice versa. They are the best for the large and sustained flows.

Proportional Meter

A proportional meter has a certain proportion of the flow passing through it and then through a displacement-type measuring device, which records the total volume. Proportional meters are good for large flows; however, they are expensive and hard to repair.

Compound Meter

It is a combination of a displacement and a current meter. A small flow is measured by the displacement mechanism and a large flow by the current mechanism. They are used for variable flows.

FIRE HYDRANTS

Hydrants provide adequate water flow for firefighting and flushing the lines. They are conveniently located for adequate firefighting and flushing the system. A fire hydrant has at least two 2.5-inch outlets and a gate valve between itself and the water main. It should be able to deliver 600 gpm with a head loss not over 2.5 pounds in the hydrant and 5 pounds from the main to the outlet. There should be a good drainage of water around it after the fire hydrant valve is closed.

WATER QUALITY IN THE DISTRIBUTION SYSTEM

It is important to maintain the proper water quality in the distribution system until the last customer is served. Water quality deteriorates due to stagnancy, dead ends, cross connections, and corrosion.

Stagnancy

Stagnancy results when water stays too long in the lines. The water loses some of the residual disinfectant and dissolved oxygen. When water is stagnant, microbes such as iron, sulfur, nitrogen, or even coliform bacteria begin to multiply and produce objectionable byproducts.

Bacteria occur in a thin layer of organic and inorganic deposition, called *biofilm,* on the inner surface of water lines, where they multiply. Part of this film can be eroded by certain hydraulic conditions, such as high velocity and water hammer to release these bacteria, resulting in high heterotrophic plate counts and even a positive coliform test. The best remedy, to stop the regrowth of microbes, is the adequate unidirectional (in the same direction as the normal flow) flushing of the distribution system and continuously maintaining an appropriate amount of residual disinfectant. Unidirectional flushing is important to maintain proper water quality by removing biofilm, maintaining disinfectant residual, and reducing erosion.

If the water pH is low and chloroamines are the residual disinfectant, then nitrification bacteria convert ammonia to nitrites, and nitrites to nitrates (nitrification).

$$\text{Ammonia} \rightarrow \text{Nitrites (NO}_2^-) \rightarrow \text{Nitrates (NO}_3^-)$$
$$\text{Nitrification}$$

To stop nitrification, the pH of water needs to be raised above 9.0. Water should be kept moving and lines should be kept flushed regularly.

Dead Ends

Dead ends are the endings of the pipes without any movement of the water. There should be a regular flushing of the dead ends and cleaning as required to keep the water fresh.

Cross Connection

Cross connection is any physical connection between the treated and non-treated water. Examples of cross connections are a direct pipe connection from a private water supply (well water) to the municipal water supply; discharge of the potable water supply below the water level in a swimming pool; and submerged inlets in lavatories, bath tubs, fountains, and spray tanks. When there is a high demand of water and pressure is low in the water lines, the nontreated water is sucked into the public water supply system, sometimes resulting in serious health problems.

Cross connection contaminations of drinking water with pesticides are reported. Sprinkler systems have also caused cross connections. There should always be backflow prevention, such as an air gap or a check valve, to protect the public water supply.

Corrosion

Corrosion of water lines and plumbing fixtures can cause the leaching of harmful metals such as lead and copper into water, growth of objectionable microbes, and biofilm formation. Water should be noncorrosive.

It does not matter how properly the water is treated unless it is delivered as good as it is treated. For proper delivery, proper maintenance of system and pressure are important.

Table 13-1 Transmission System Problems and Their Solutions

Problems	Possible Causes	Possible Solutions
Water smells like a pesticide.	There is a cross connection. If pesticide is being sprayed in an area, look for the water line in the tank. A hose under the water in the tank is causing the problem.	Remove the hose and flush the area. Check the reservoir for any contamination. If contaminated, flush it, and clean it thoroughly before putting it back in service.
High heterotrophic bacterial count in the distribution system when chloroamines are used for postdisinfection.	Nitrification, which is the conversion of ammonia into nitrites and then into nitrates by bacteria. It is indicated by pH around 8, stagnant smell of water, high nitrites and nitrates and red water complaints.	Increase the pH to an appropriate level (about 9) to solve the nitrification problem.
	Dead ends. This is indicated by red and stagnant water complaints and high HPC count. It occurs when chlorine gets low and dissolved oxygen is depleted in the stagnant water. Iron and sulfur bacteria become active and corrosion of pipes starts.	Unidirectional (in one direction throughout the system to prevent the erosion of deposits in the pipe) flushing of the area is the solution to this problem.
Coliform bacteria are in the distribution system.	Contaminated sample.	Make sure that all equipment is properly sterilized; there is no contamination of the sample or the equipment while collecting or testing the sample. Carefully, resample the site, test by two different methods (colilert and membrane filter) to reconfirm the results.
	Breakdown of treatment.	Check all bacteriological test results of the treatment train; be sure the water is properly disinfected. Increase the disinfectant residual in the distribution system.

Symptom	Possible cause	Solution
	Change in the direction of water flow in the lines.	It will cause the erosion of the biofilm. Correct all flows to unidirectional.
	Sewage cross connection.	Check for any water or sewer line breaks in the area. Correct the situation, and flush the area.
	Regrowth of coliform in biofilm.	Biofilm has certain coliform bacteria which start growing as soon as conditions are favorable, such as low disinfectant residual and low dissolved oxygen. Flush the area thoroughly, increase disinfectant residual, and check for coliforms. It may need the pigging of lines to remove biofilm. Consistently, maintain the clean system and proper disinfectant residual.
Pinkish deposits are on faucets.	High calcium carbonate in the water.	The color is due to a fungus growing on calcium carbonate deposits. Adjust the alkalinity and pH to reduce the calcium carbonate deposition potential.
There is a high chlorite content.	Inadequate ferrous treatment. Ferrous should be added before alum or lime.	Determine the needed dose of ferrous to remove the desired amount of chlorite. Apply three parts of ferrous for each part of chlorite.
There is a high chlorate content.	An old sodium hypochlorite solution used for chlorination can have high chlorates.	Always use a fresh sodium hypochlorite solution.
	Concentrated sodium hypochlorite has a high amount of chlorates.	Normally sodium hypochlorite is 15%, dilute it to 7.5%.

TRANSMISSION SYSTEM PROBLEMS AND POSSIBLE SOLUTIONS

Refer to Table 13-1 for common transmission system problems.

QUESTIONS

1. What is a pump? Explain the term *centrifugal force* and its application in the operation of a centrifugal pump. What is the major difference between a centrifugal and a displacement pump?

2. Explain these terms:
   a. Suction head
   b. Suction lift
   c. Total static head
   d. Total dynamic head

3. Define the following terms:
   a. Impeller
   b. Axial flow
   c. Radial flow
   d. Francis flow

4. What is a pump curve?

5. What are the two main factors of a centrifugal pump that control its capacity?

6. What is the practical suction lift of a centrifugal pump? Explain the term *shut off* head of a pump.

7. What is the effect of a closed discharge valve on a plunger pump?

8. Write two important advantages of a centrifugal pump over a displacement pump.

9. a. What are the main differences between a gate valve and a butterfly valve?
   b. What kind of valve is used to control the water level in a reservoir?
   c. What can happen if a needle valve is fully open?

10. What is a cross connection, and how can it be prevented?

14

HYDRAULICS

Hydraulics, the Greek word for water pipe, is the science of fluids, such as water. For water treatment, it is important to understand the pressure and the flow behavior of water. The study of water at rest and in motion is known as *hydrostatics* and *hydrodynamics*, respectively. Hydraulics explains the behavior of water at rest (as in elevated tanks) and in motion, while flowing through pipes, channels, and pumps.

FORCE AND PRESSURE

Force

Force (F) is the weight on the bottom of a column of water (e.g., 1 cubic foot of water exerts 62.4 lb. of force on 1 ft.^{2} of its bottom).

1 cubic ft. of water = 7.48 gal./ft.^{3} × 8.34 lb./gal. = 62.4 lb.

Thus, force at the bottom of one cubic foot of water = 62.4 lb./ft.^{2}

Pressure

Pressure (P) is the force per unit area.

$$P = F/A \text{ and } F = P \times A$$

where:

P = pressure
F = force
A = area

Using 1 ft.^{3} of water

F = 62.4 lb./ft.^{2}
P = (62.4 lb./ft.^{2})/(144 in.^{2}/ft.^{2})= 0.433 lb./in.^{2}

    Therefore, 1 ft. column of water exerts a pressure of 0.433 pounds/in^2, psi
And 1 psi = 1 ft./0.433 psi = 2.31 ft. (height of water)
    Thus, we can convert the feet of water column, commonly called *head,*
into psi, and vice versa.

Example: If a force of 15 lb. is applied at point #1 with an area of 2 in.^{2},
what will the lifting force at point #2 with an area of 20 in.^{2} (see Figure
14-1)?

$$\text{Force at point \#2} = (15 \text{ lb.}/2 \text{ in.}^2) \times 20 \text{ in.}^2 = 150 \text{ lb.}$$

    Static pressure, or static head, is the vertical distance between any base
point (like ground level, the top of a hill, or the bottom of a valley) and the
free surface of the water (see Figure 14-2). Free surface level of intercon-
nected tanks and lines of a confined liquid is always the same. It is explained
by Evangellista Torricelli's law, which states that *a jet of water from the
bottom of the source will rise to the level of its source if directed upwards,
as with a flowing artesian well.*
    Water at rest with free surface has the same pressure at all points at the
same level. For example, all homes at the same level will have the same
pressure. *Pascal's law* states that a *pressure exerted on a liquid in a closed
system is transmitted undiminished in all directions at the same force at right
angles to the surface.* See Figure 14-3.

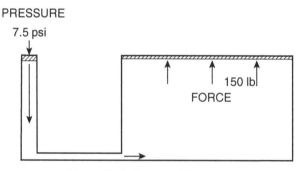

Figure 14-1 Force and Pressure

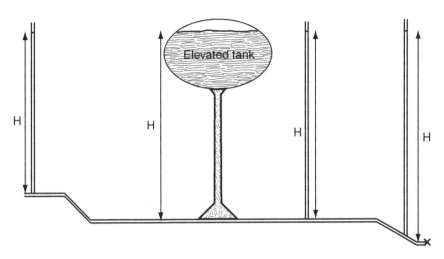

Figure 14-2   Static Head at Different Locations around an Elevated Tank

Therefore, a confined liquid exerts equal pressure perpendicular to all parts of the inner surface of its container. Liquid will be forced through the bottom or a side hole of the container with the same pressure when the pressure is applied at the top (e.g., a piston pump). Standard (at sea level) atmospheric pressure is 14.7 psi, which forces water up when suction is applied (siphoning) at the upper end of a pipe with its lower end below the water surface. Functioning of a pump is based on this principle; suction is applied at the suction end of the pump to suck water to the discharge end. We apply this principle when we use a straw to suck a drink from a glass. Standard atmospheric pressure is 34 feet (14.7 psi × 2.31 ft/psi) water head; therefore, 34 feet is the theoretical height to which water can be lifted. It is called the *theoretical lift* of a pump.

Velocity head (V_h) is the pressure due to velocity, the motion of water.

Daniel Bernoulli's law states that the *higher the velocity of a liquid, the lower is the pressure,* and vice versa. When a pipe narrows, velocity of the

10 psi

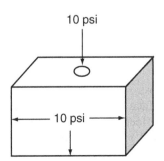

10 psi

Figure 14-3   Pascal's Law

flowing liquid increases and its pressure decreases. Due to resistance to flow, the pressure ahead of the restriction increases. For example, our blood pressure increases when our arteries become narrower.

$$PV = P_1V_1$$

where:

P = pressure corresponding to velocity V
P_1 = pressure corresponding to velocity V_1

This principle is applied in the designing of a centrifugal pump, where velocity is converted into pressure.

Water standing in an elevated tank has a high pressure and no velocity. When a discharge valve at the bottom of a tank is opened, the velocity increases and the pressure falls.

Torricelli's law states that the velocity of a jet of water coming through an opening is equal to the velocity of a falling body from the air from the surface of the water to the opening. Thus, the falling water follows the law of gravity, like any other falling object; it causes the gravity flow.

Water flow due to gravity from a higher to a lower elevation is known as *gravity flow*. The rate of velocity corresponds to the elevation differential. The more the difference in two elevations, the higher is the velocity, and vice versa. The following equation converts the velocity to velocity head:

$$V_h = V^2/2g$$

where:

V_h = Velocity head
g = 32.2 ft./sec./sec., the acceleration due to gravity
V = velocity

Example: Suppose water is moving in a line with a velocity of 10 ft./sec. What is its velocity head?

$$V_h = (10 \text{ ft./sec.} \times 10 \text{ ft./sec.})/(2 \times 32.2 \text{ ft.})$$

$$\text{Velocity head} = 1.5 \text{ ft.}$$

Friction head is the loss of pressure due to friction (resistance to flow) between water and the surface of a pipe. It is negative and generally expressed as the head loss per 1,000 feet of pipe. The larger the diameter of a pipe, the less is the friction loss, and vice versa. Friction loss is directly proportional

to the length, velocity, and the roughness of a pipe. The roughness is represented by C factor that varies from 155 for very smooth and large pipes to
as low as 40 or even less for badly corroded or rough pipes. For a normal
clean and smooth pipe, the factor is 100. Mostly, we use a nomogram for
friction loss developed from Hazen–Williams formula with C value of 100.
This nomogram gives the relationship of flow, friction loss, and the pipe
diameter (size). Figure 14-4 shows these relationships.

Example: An elevated tank, on a 100-foot-high stand pipe, is 50 feet deep.
It has three customers on a 4 inch line with a 90 gpm flow. What will be the
pressure for each of them if customer #1 is 2,000 feet away from the elevated
tank, customer #2 is 1,000 feet away from the first, and customer #3 is 1,500
feet away from the second. The first two are at the baseline ground level, and
the third is living on a 50 ft high hill (above the base line). See Figure 14-5.

Using the Figure 14-4 for friction losses, the pressure at each of the three
different sites will be:

Pressure for customer #1
= (150 ft − 20 ft. friction loss for 2,000 ft.)/2.31 ft/psi
= 56 psi

Pressure for customer #2
= psi at customer #1 − friction loss for another 1,000 ft., which is 10 ft.
= 56 psi− (10 ft./2.31 ft./psi)
= 56 psi − 4.33 psi
= 51.7 psi

Pressure for customer #3
= 51.7 psi − (hill height + friction loss for 1,500 ft.)
= (51.7 psi) − {(50 ft. + 15 ft.)}/2.31 ft./psi
= 51.7 psi − 28.14 psi = 23.56 psi

Normally, 40 psi is the minimum required pressure for the fire protection.
Thus, the pressure is too low for the third customer. This example illustrates
the importance of the proper size of lines and elevations in distribution system. To maintain a proper pressure in the system, it is important for the water
utility to know the highest and the lowest elevation sites in its distribution
system.

FLOW

Flow is the quantity of water flowing per unit time. Mostly, it is expressed
as cubic feet per second (cfs) for a river or stream and gallons per minute

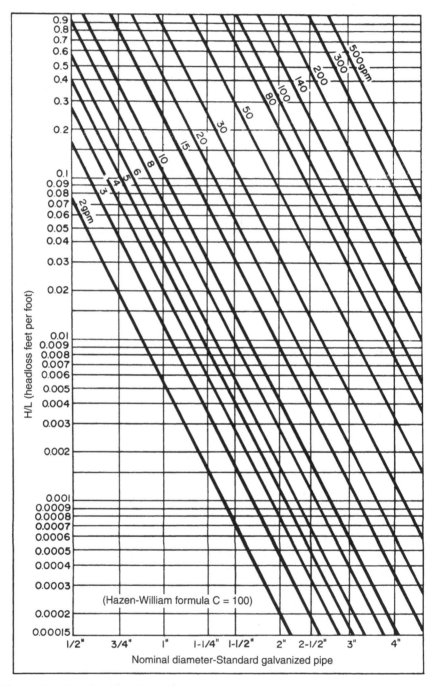

Figure 14-4 Relationship of Flow, Diameter, and Friction Less, based on Hazen-Williams Formula.

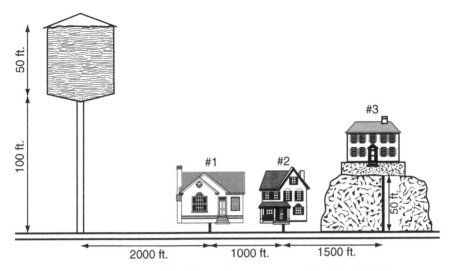

Figure 14-5 Pressure for Different Customers around an Elevated Tank

(gpm) or million gallons per day (mgd) for water flow to and from the treatment plant. Flow is directly proportional to the square of the diameter or the radius of a pipe.

$$Q/Q_1 = r^2/r_1^2$$

where:

Q = flow for radius r
Q_1 = flow for radius r_1

Suppose, we compare the capacities of 2- and 4-inch pipes. If flow through a 2-inch pipe is 5 cfs, the flow of a 4-inch pipe will be 20 cfs, which is four times more.

$$5 \text{ cfs}/Q_1 = 1^2/2^2 \text{ or } Q_1 = 5 \text{ cfs} \times 4 = 20 \text{ cfs}$$

Generally, it is beneficial to install larger pipes than smaller ones.

WATER HAMMER

Water hammer is the surge or thrust of pressure, caused by a sudden change in the flow (velocity) by an abrupt opening or closing of a valve or a fire hydrant, or an abrupt stopping and starting of a pump. Water hammer causes high pressure due to surge in the pipes, which can damage the joints at the nearest fitting of the line. Besides high pressure, it dislodges sediments in the lines, resulting in the dirty water or red water complaints. Gradual opening and closing of valves and fire hydrants, and less frequent starting and stopping of pumps, reduce the impact of water hammer. Generally, the fittings of the water pipes are strong enough and well protected by anti-water-hammer material around them, such as thrust blocks and joint restraints.

QUESTIONS

1. Define the terms *hydraulics, force, pressure, static head,* and *friction head.*

2. The water surface in an elevated tank is 139 feet high from the ground level. What is the pressure as psi at a house close to the tank?

3. **a.** Friction loss or friction head is less in wider pipes. T or F
   b. Small pipes with very rough insides have more or less friction loss?

4. **a.** From Figure 14-4, determine the friction head for 500 feet of 3-inch ordinary smooth pipe at 200 gal/minute flow.

5. Using Figure 4-14 find the friction loss for 1, 2, and 4 inch pipes for 20 gal/min flow.

6. **a.** Show that 2-inch pipe has four times more capacity than 1-inch pipe.
   b. The higher the elevation of a house, the lower is the water pressure, and vice versa. T or F
   c. Higher the velocity, smaller the diameter, rougher the inner surface, the less is the pressure. T or F

7. A customer is 20 feet above the ground level and 1,000 feet away from a 120-foot high elevated tank. What will be the pressure for 3 inch pipe and 200 gpm flow?

8. Explain the term *water hammer* and the measures to avoid its excessive force.

15

MATHEMATICS

Mathematics is the study of numbers, lines, surfaces, volumes, and flow rates. For calculations at different phases of water treatment, an understanding of math is needed to determine areas, volumes, detention times, flow rates, and chemical feed rates.

METRIC UNITS OF MEASUREMENTS

- *Gram* is the basic unit of mass. One gram is the mass of one cubic centimeter (cc) of water at 4°C. One gram is equal to 0.0022 pounds (lb.).
- *Meter,* the basic unit of length, is defined as 1/10,000,000 of the distance to the North Pole from the equator in the Paris Meridian. One meter is equal to 39.37 inches.
- *Liter* is the basic unit of volume. It is the volume of one kilogram of water at 4°C. One liter is equal to 0.264 gallons.

Multiple and fractional parts of these units have simple numerical relationship based on number 10. They are assigned the Greek prefixes for multiples and the Latin prefixes for fractions as follows:

Deka, D = 10, or (10^1) basic units

Hecto, H = 100, or (10^2) basic units

Kilo, K = 1,000, or (10^3) basic units

Mega, M = 1,000,000, or (10^6) basic units

Giga, G = 1,000,000,000, or (10^9) basic units

Tera, T = 1,000,000,000,000, or (10^{12}) basic units

Deci, d = 1/10, or (10^{-1}) of basic unit

Centi, c = 1/100, or (10^{-2}) of basic unit

Milli, m = 1/1,000, or (10^{-3}) of basic unit

Micro, μ = 1/1,000,000, or (10^{-6}) of basic unit

Nano, n = 1/1,000,000,000, or (10^{-9}) of basic unit

Pico, p = 1/1,000,000,000,000, or (10^{-12}) of basic unit

Femto, f = 1/1,000,000,000,000,000, or (10^{-15}) of basic unit

Atto, a = 1/1,000,000,000,000,000,000, or (10^{-18}) of basic unit

By using meter as an example, these prefixes are used as follows:

Kilometer (Km) = 1,000 meters = 100 decameters = 10 hectometers

Hectometer (Hm) = 100 meters

Dekameter (Dm) = 10 meters

Meter

Decimeter (dm) = 1/10 meter

Centimeter (cm) = 1/100 meter

Millimeter (mm) = 1/1,000 meter = 1/100 decimeter = 1/10 centimeter

Temperature in metric system is measured on the centigrade (Celsius) scale, which ranges from 0 to 100°. Zero degree is the freezing point of water, and 100° is the boiling point. For practical purposes, the Fahrenheit scale is used, which has 32° as the freezing point and 212° as the boiling of water.

The following equations convert the temperature from one scale to the other:

$$°F = (°C \times 9/5) + 32$$
$$°C = (°F - 32) \times 5/9$$

Example. Convert 95°F to °C and 20°C to °F.

$$95°F = (95 - 32) \times 5/9 = 35°C$$
$$20°C = (20 \times 9/5) + 32 = 68°F$$

SCIENTIFIC OR EXPONENTIAL NOTATION

Scientific notation is a logarithmic shorthand method of writing very large and very small numbers, such as 2,000 pounds/ton and 0.002205 pounds/gram.

$$\text{Scientific notation} = M \times 10^n$$

where:

M = a number from 1.0 to 9.99
n = exponent or the power of the base 10, known as log

n is positive for the numbers higher than 10 and negative for the numbers less than one. The following examples show the system:

$$2,000 \text{ pounds/ton} = 2.0 \times 10^3 \text{ pounds/ton}$$

$$0.002205 \text{ pounds/gram} = 2.205 \times 10^{-3} \text{ pounds/gram}$$

Power of base 10 is the number of places the decimal point is moved to have one digit to the left. If decimal point is moved to the left, as in the case of pounds/ton, power is positive; if to the right, as in the second example, it is negative.

IMPORTANT CONVERSION FACTORS FOR WATER TREATMENT

- 1 gallon of water = 8.34 pounds (lb.)
- 1 cubic foot (ft.^{3}) = 7.48 gallons (gal.)
- 1 pound = 7,000 grains (gr.)
- 1 horsepower (HP) = 33,000 lb./min./ft. = 33,000 lb./8.34 lb./gal./min./ft. = 3,957 gal./min./ft.
- 1 horsepower = 0.746 kilowatts (Kw)

Most conversion factors can be derived from these five:

- 1 ft. head of water = 0.433 lb./in.^{2} (psi)
  It is the pressure on 1 in.^{2} area at the bottom of one cubic foot of water. 1 ft. head of water = (7.48 gal./ft.^{3} × 8.34 lb./gal.)/144 in.^{2} = 0.433 lb./in.^{2}
- 1 psi = 2.31 ft. of water
  1 psi = 1/0.433 ft. = 2.31 ft.

- 1 cubic feet per second (cfs) = 0.65 million gal./day (mgd)
  1 cfs = (7.48 gal./ft.^{3}) × {(60 sec./min.) × 1,440 min./day)}/1,000,000
  = 0.65 mgd
- 1 mgd = 1/0.65 cfs = 1.55 cfs
- 1 mgd = 694 gpm
  1 mgd = 1,000,000 gal./day)/(1,440 min./day) = 694 gpm
- 1 part per million parts (ppm) = 8.34 lb./million gallons (MG) of water
  1 gal./1,000,000 gal. of water = 8.34 lb./MG of water
- 1 mg/L = 1 part per million (ppm)
  1 mg/Kg (wt. of 1 liter of water) = 1 mg/1,000,000 mg = 1 ppm
- 1 gr./gal. = 143 lb./MG
  1 gr./gal. = (1,000,000 gr./7,000 gr./lb.)/MG = 143 lb./MG
- 1 gr./gal = 17.1 ppm
  1 gr./gal. = (143 lb./MG)/(8.34 lb./MG/ppm) = 17.1 ppm

AREAS

Area of an object is the measurement of its surface. Surface is the product of two dimensions, using the same linear units, such as inches times inches, feet times feet, and yards times yards. Therefore, areas are measured as square inches (in.^{2}), square feet (ft.^{2}), and square yards (yd.^{2}). The two basic shapes of our main concern are rectangles and circles (see Figure 15-1). From these, other areas can be found.

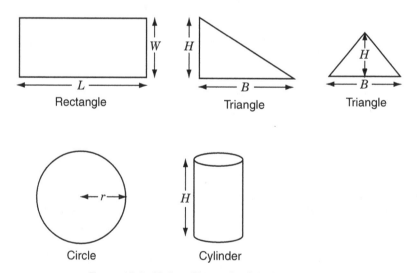

Figure 15-1   Various Shapes for Calculating Areas

Area of a Rectangle

Area of a rectangle = length × width.

$$A = L \times W$$

where:

A = area of a rectangle
L = length
W = width

Example. Calculate the area of the base of a 100-foot-long and 50-foot-wide rectangular tank.

$$\text{Area} = 100 \text{ ft.} \times 50 \text{ ft.}$$
$$= 500 \text{ ft.}^2$$

Area of a Triangle

A triangle is 1/2 of a rectangle; therefore, the area of a triangle is 1/2 the area of a rectangle. The base and height of a triangle is the same as the length and width of the rectangle.

$$A = 1/2 \, B \times H$$

where:

A = area of a triangle
B = base
H = height

Example. Calculate the area of a triangle with a 10-foot base and a 7-foot height.

$$\text{Area} = 1/2(10 \text{ ft.} \times 7 \text{ ft.})$$
$$= 35 \text{ ft.}^2$$

Area of a Circle

$$\text{Area of a circle} = \pi r^2$$

where:

π = 22/7 or 3.14. It is the Greek letter pi, which stands for a factor, the multiplier of diameter of a circle, to obtain the circumference (πd). It cannot be written as an exact decimal fraction; therefore, it is called an irrational number.

$$r = \text{radius} = 1/2 \text{ of the diameter } (d)$$

This formula can be simplified into $0.785d^2$ by using the following equation:

$$\text{Area of a circle} = 3.14 \times 1/2d \times 1/2d$$
$$= (3.14/4) \times d^2$$
$$= 0.785 \times d^2$$

Example. Calculate the area of the top lid of a cylindrical tank of 20-foot diameter.

$$\text{Area} = 3.14 \ r^2 = 3.14 \times 10 \text{ ft.} \times 10 \text{ ft.} = 314 \text{ ft.}^2$$

By using the formula, $0.785d^2$

$$\text{Area} = 0.785 \times 20 \text{ ft} \times 20 \text{ ft.} = 314 \text{ ft.}^2$$

Area of a Sphere

The area of four circles forms the surface of a sphere.

$$\text{Area of a sphere} = 4 \times 3.14 \ r^2 = 4 \times 0.785d^2 = 3.14 \ d^2$$

where:

r = radius
d = diameter

Example. Calculate the area of a spherical water tower, 50 feet in diameter.

$$\text{Area} = 3.14d^2$$
$$= 3.14 \times 50 \text{ ft.} \times 50 \text{ ft.} = 7,850 \text{ ft.}^2$$

Area of a Cylinder

A cylinder is a rectangle rolled into a cylinder by holding one side of the rectangle adjacent to the other parallel side. Thus, circumference becomes one side and the height the other side of a rectangle. Therefore, lateral area of a cylinder is the circumference multiplied by the height.

Lateral area of a cylinder = $2\pi r$, (circumference) $\times H$ = $2 \times 3.14 \times r \times H$ = $6.28 \times r \times H$

where:

r = radius
H = height

Example. Calculate lateral area of a cylindrical chemical storage tank, 20 feet in diameter and 20 feet high.

$$Area = 6.28 \times 10 \text{ ft.} \times 20 \text{ ft.} = 1,256 \text{ ft.}^2$$

Total area of a cylinder is the sum of its lateral area, top area, and base area.

$$Lateral \text{ area} = 2\pi r \times H$$
$$Top \text{ and base areas} = 2 (2\pi r^2)$$
$$Total \text{ area} = (2\pi r \times H) + 2 (2\pi r^2)$$

It can be simplified into the following:
Total area = $2\pi r (H + r)$ because $2\pi r$ is the common factor in the formula.

Example. Calculate the total area of a cylinder, 20 ft in diameter and 20 feet high.

$$Area = 2 \times 3.14 \times 10 \text{ ft.} (20 \text{ ft.} + 10 \text{ ft.}) = 1,884 \text{ ft.}^2$$

VOLUMES

The volume of an object is the space it occupies. Volume has the third dimension, which is height or depth. It is the area of the base of the object multiplied by its height or depth. All units in these three parameters must be the same—that is, inches only, feet only, or yards only. The volume of an object is expressed in cubic inches (in.^{3}), cubic feet (ft.^{3}), and cubic yards (yd.^{3}) (see Figure 15-2). These units are converted into gallons by using the appropriate conversion factor. For example, cubic feet multiplied by 7.48 gal./ft.^{3} become gallons.

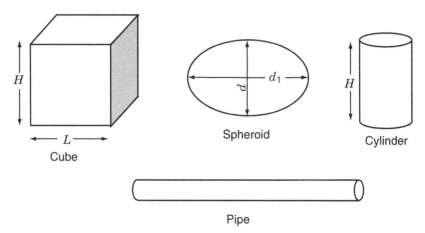

Figure 15-2 Various Objects for Calculating Volumes

Volume of a Rectangular Tank

A rectangular object with depth becomes a rectangular tank.

$$\text{Volume of a rectangular tank} = L \times W \times H \text{ or } D$$

where:

L = length
W = width
H = height
D = depth

Example. Calculate the volume of a rectangular tank, 100 feet long, 50 feet wide, and 15 feet deep. Convert the volume into gallons.

$$\text{Volume} = 100 \text{ ft.} \times 50 \text{ ft.} \times 15 \text{ ft.} = 75,000 \text{ ft.}^3$$
$$= 75,000 \text{ ft.}^3 \times 7.48 \text{ gal.}/\text{ft.}^3$$
$$= 561,000 \text{ gallons}$$

Volume of a Cylinder

The volume of a cylinder is the area of the circular base multiplied by the height of the cylinder.

$$\text{Volume} = \text{Area of the base} \times \text{Height or Depth} = 3.14 \, r^2 \times H \text{ or } D$$

This formula can be written as

$$\text{Volume of a cylinder} = 0.785 \times d^2 \times H \text{ or } D$$

where:

r = radius
d = diameter
H = height
D = depth

Example. Calculate the volume of a cylindrical tank, 100 feet in diameter and 20 feet deep.

$$\text{Volume} = 0.785 \times d^2 \times D = 0.785 \times 100 \text{ ft.} \times 100 \text{ ft.} \times 20 \text{ ft.}$$
$$= 157,000 \text{ ft.}^3 = 157,000 \text{ ft.}^3 \times 7.48 \text{ gal./ft.}^3$$
$$= 1,174,360 \text{ gallons}$$

Volume of a Cone

A cone is one-third of a cylinder; thus, its volume is one-third the volume of a cylinder with the same base and height.

$$\text{Volume of a cone} = 1/3 \times 0.785 \times d^2 \times H = 0.262 \times d^2 \times H$$

Example. Calculate the volume of a conical sludge hopper at the base of a circular sedimentation basin. The hopper is 20 feet in diameter and 10 feet deep.

$$\text{Volume of the hopper} = 0.262 \times 20 \text{ ft.} \times 20 \text{ ft.} \times 10 \text{ ft.}$$
$$= 1,047 \text{ ft.}^3 = 1,047 \text{ ft.}^3 \times 7.48 \text{ gallons/ft.}^3$$
$$= 7,832 \text{ gallons}$$

Volume of a Sphere

$$\text{Volume of a sphere} = 4/3\pi r^3$$
$$\text{Simplified formula} = 4/3 \times (3.14 \times d/2 \times d/2 \times d/2) = 0.5233 \times d^3$$

Example. If a spherical elevated tank is 50 feet in diameter, what is its capacity in gallons?

$$\text{Volume of the tank} = 0.5233 \times d^3 = 0.5233 \times 25 \text{ ft.} \times 25 \text{ ft.} \times 25 \text{ ft.}$$
$$= 8,176.6 \text{ ft.}^3 \times 7.48 \text{ gal./ft.}^3 = 61,161 \text{ gal.}$$

Volume of a Spheroid Tank (Oval)

Volume of a spherical tank $= (\pi/6) \times (d^2 \times d_1)$

Simplified formula $= (3.14/6) \times d^2 \times d_1 = 0.5233 \times d^2 \times d_1$

where:

d = small diameter
d_1 = long diameter

Example. Calculate the capacity of a spheroid tank of 20 feet and 30 feet diameters.

Volume $= 0.5233 \times d^2 \times d_1 = 0.5233 \times 20$ ft. $\times$ 20 ft. $\times$ 30 ft.

$= 6,280$ ft.$^3 = 6,280$ ft.$^3 \times 7.48$ gal./ft.$^3 = 46,971$ gal.

Volume of a Pyramid Shape Hopper

Like a cone, it is one-third of the rectangular tank with the base of the pyramid.

Volume of a pyramid $= 1/3 \times L \times W \times H$

Example. What is the volume in gallons of a hopper, 100 feet long, 50 feet wide, and 15 feet deep?

Volume $= 1/3 \times 100$ ft. $\times$ 50 ft. $\times$ 15 ft. $= 25,000$ ft.^{3}

$= 25,000$ ft.$^3 \times 7.48$ gal./ft.$^3 = 187,000$ gallons

Volume of a Pipe

A pipe is a long cylinder; thus, its volume is calculated like that of a cylinder. Usually, the length is in miles, which must be converted into feet for the calculation.

Volume of a pipe $= \pi r^2 \times L = 0.785 \times d^2 \times L$

where:

r = radius, in ft.
d = diameter, in ft.
L = length, in ft.

Example. A water line is 48 inches in diameter (commonly called a 48-inch line) and 1.5 miles long. Calculate its volume in gallons.

$$\text{Volume} = 0.785 \times 4 \text{ ft.} \times 4 \text{ ft.} \times 1.5 \text{ miles} \times 5{,}280 \text{ ft./mile}$$
$$= 99{,}475 \text{ ft.}^3 = 99{,}475 \text{ ft.}^3 \times 7.48 \text{ gal/ft.}^3 = 744{,}074 \text{ gallons}$$

Detention Time

Detention time is the time it takes to fill a tank or a pipe.

$$\text{Detention time } (DT) = V/Q$$

where:

V = volume, as ft.^{3} or gallons
Q = flow rate, as ft.^{3}/sec. (cfs) or gal./min. (gpm).

Example. Calculate the detention time of a tank 100 feet in diameter, 20 ft. deep, and at a flow of 10 mgd.

$$V = 0.785 \times 100 \text{ ft.} \times 100 \text{ ft.} \times 20 \text{ ft.} = 157{,}000 \text{ ft.}^3$$
$$Q = 10 \text{ mgd} \times 1.55 \text{ cfs/mgd}$$
$$= 15.5 \text{ cfs}$$
$$DT = V/Q = 157{,}000 \text{ ft.}^3/15.5 \text{ cfs} = 10{,}129 \text{ seconds}$$
$$= 10129 \text{ sec./}3600 \text{ sec./hr.}$$
$$= 2.8 \text{ hours}$$

Using mgd,

$$DT = (157{,}000 \text{ ft.}^3 \times 7.48 \text{ gal./ft.}^3)/10{,}000{,}000 \text{ gal./day}$$
$$= 1{,}174{,}360 \text{ gal./}10{,}000{,}000 \text{ gal./day}$$
$$= 0.1174360 \text{ days} \times 24 \text{ hrs./day}$$
$$= 2.8 \text{ hours}$$

Flow

Flow or flow rate is the cfs, gpm, or mgd flow through a tank or a pipe (see Figure 15-3).

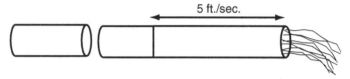

Figure 15-3 A Flowing Pipe

$$Q = V \times A$$

where:

Q = flow as cfs
V = velocity as ft./sec.
A = cross-sectional area of the tank or pipe, as ft.^{2}

Example. Calculate the flow rate for a 6-inch pipe with 5 ft./sec. velocity. It is the volume of 5 feet in length of 6-inch pipe, emptying per second.

Q = 3.14 × 0.25 ft. × 0.25 ft. × 5 ft./sec. = 0.98 ft.^{3}/sec.

= (0.98 ft.^{3} × 7.48 gal./ft.^{3}/sec.) × 60 sec./min. = 440 gal./min.

Weir Overflow Rate

Weirs are rectangular or V-notch openings used for the overflow rate of water from the sedimentation basins.

$$\text{Overflow} = (\text{gal.}/\text{day})/\text{ft. of weir}$$

Example. Calculate the weir overflow rate in gal./ft./min. for a 1,000-foot-long weir and 1 mgd flow.

Weir overflow = (1,000,000 gal./day)/1,000 ft. = 1,000 gal./ft./day

= (1,000 gal./ft./day)/1,440 min./day = 0.7 gal./ft./min.

Surface Loading or Overflow Rate

It is the overflow rate for basins and filters as gal./day/unit surface area of a tank or a filter.

$$\text{Surface loading} = (\text{Flow}/\text{day})/\text{Surface area, such as ft.}^2$$

Example. A primary sedimentation basin 50 feet long and 20 feet wide has a flow rate of 2 mgd. What will be its surface loading as gal./ft.^{2}/day?

$$\text{Surface loading} = (2{,}000{,}000 \text{ gal./day})/(50 \text{ ft.} \times 20 \text{ ft.})$$
$$= 2{,}000 \text{ gal./ft.}^2/\text{day}$$

CHEMICAL DOSES OR FEED RATES

Feed rates are based on the fact that one gallon of water weighs 8.34 pounds. 1 gallon in 1 million gallons of water is 1 part per million (ppm); thus, in practice, 8.34 pounds of most of the chemicals/million gallons (MG) of water are considered 1 ppm, regardless of the density of the chemical.

$$1 \text{ ppm of a chemical} = 8.34 \text{ lb. of the chemical/MG}$$

Dose as ppm

Pounds of a chemical feed required per day = (8.34 × ppm dose × flow as mgd)/% purity of the chemical in decimal (70% will be 0.70).

$$\text{ppm, dose} = (\text{Pounds of chemical used/day})/(\text{mgd} \times 8.34)$$

Example.

1. How many pounds of 70 percent chlorine powder are required per day to treat 20 mgd flow and 4 ppm dose?

$$\text{Pounds of 70\% chlorine/day} = (8.34 \times 4 \times 20)/0.70$$
$$= 953 \text{ lb.}$$

2. How many ppm of lime dose will be if we used 2,000 pounds of lime/day to treat 10 mgd flow?

$$\text{Lime dose} = 2{,}000 \text{ lb.}/(10 \text{ mgd} \times 8.34 \text{ lb./MG}) = 24 \text{ ppm}$$

Dose as Grains per Gallon

Grains/gallon dose is calculated by using the conversion factor 143 pounds per million gallons of water equals 1 grain per gallon (grpg) dose.

Pounds per day = 143 × grpg × mgd

And, grpg dose = (Pounds/day)/(143 × mgd)

Example. Calculate the lime usage as pounds per day for treating 20 mgd water with 3 grpg dose.

Pounds of lime/day = (143 lb./MG/grpg) × 3 grpg × 20 mgd = 8,580 lb.

HORSE POWER

Water Horsepower (WHP)

Water horsepower is the work done by a pump by lifting 33,000 lb. or 3,957 gallons of water/ft./minute. It is known as *output.*

WHP = {(gpm, pump capacity) × head in ft.)}/3,957 gal./ft./min./HP

Example. A pump has 800 gal./min. capacity. Determine its water horsepower for 15 foot head.

WHP = {(800 gal./min.) × 15 ft.)}/3,957 gal./ft./min./HP = 3 HP

Brake Horsepower (BHP)

The brake horsepower is the amount of energy actually used by the pump to do the work. It is energy input.

BHP = WHP/pump efficiency

Efficiency = WHP/BHP

Example. If the pump in the previous example is 70 percent efficient, what will be the brake horsepower?

BHP = 3 HP/0.70 = 4.3 HP

1 horsepower = 0.746 Kilowatts (Kw)

Thus, the power used or input = 4.3 HP × 0.746 Kw/HP = 3.2 Kw

Motor Horsepower (MHP)

The motor that powers the pump is not 100 percent efficient. Therefore, MHP is power usage for BHP divided by the motor efficiency. If the motor for the pump previously described is 70 percent efficient, then the actual power usage is 6.14 HP, or 4.6 Kw.

MHP = BHP/motor efficiency

MHP of the pump = 4.3 HP/0.70 = 6.14 HP $\times$ 0.746 Kw/HP = 4.6 Kw

QUESTIONS

1. Define the units gram, liter, and meter.

2. Convert 42°C to °F, and 300°F to °C.

3. Write 3,957 gallons and 0.000,005 ton in the scientific notation form.

4. **a.** Why are 8.34 lb./million gallons equal to 1 ppm?
   b. Show that 1 gr./gal. is equal to 143 lb./million gal.

5. **a.** One gr./gal. is equal to 17.1 ppm. T or F

6. **a.** One psi is equal to 0.433 ft. head of water. T or F
   b. How many lb./ft./min. are equal to 1 HP?
   c. How many gallons are in 1 ft.^{3} of water?
   d. Area of a circle is $0.785d^2$. T or F

7. Calculate the area, volume, and detention time of a tank, 20 feet in diameter and 10 feet deep, and at 2 mgd flow.

8. What will be the detention time for a final basin 50 feet long 15 feet wide, with 10 feet deep water and 10 mgd flow?

9. A 48-inch (diameter) transmission line from presedimentation basin to the treatment plant is 1 mile long. How long will it take (detention time) for the water to reach the plant at 20 mgd flow?

10. Calculate the volume of a spherical tank, 30 feet in diameter.

11. Determine the capacity of a spheroid tank with the small diameter 10 feet and the large diameter 15 feet.

12. **a.** Determine the pounds of chlorine required to treat 5 mgd water flow with 4 mg/L dose by using pure chlorine.
    b. How many gr./gal. is this dose?

13. A water tower is 231 feet high. How much pressure does it have at the ground level?

14. A pump is pumping 10 mgd water to 10 feet in height. Calculate the water horsepower and break horsepower for this pump with 70 percent efficiency.

16

CHEMISTRY

Chemistry is the study of composition and changes in composition of matter. All matter exists in three physical states: solid, liquid, or gas. Water exists, simultaneously, in these three states as ice, water, and vapor. Substances react to form new substances. For example, hydrogen and oxygen react to form water. These changes are called chemical changes. Chemistry is important to understand chemical processes, such as coagulation, softening, corrosion control, and disinfection.

ELEMENTS

All matter is formed of basic substances called *elements,* their mixtures, and compounds. An element does not change chemically in normal chemical reactions. For example, the elements hydrogen and oxygen combine to form water, and water can be changed back into hydrogen and oxygen. There are 112 known elements, 92 of which are natural. Some common elements are hydrogen, oxygen, nitrogen, chlorine, calcium, magnesium, iron, and lead.

Symbol

Each element is assigned a symbol, to write its lengthy name in a short-hand form. A symbol is either the first letter of the name of an element, or it is the first and the second well-pronounced letter from the name if two or more elements start with the same letter (e.g., carbon, chlorine, calcium, cobalt, and copper). The first letter of symbol is capitalized, and the second is lowercase. Some symbols come from foreign languages like Latin, Greek, and German. For example, symbols for copper, lead, and iron are Cu, Pb, and Fe, which come from their respective Latin names cuprum, plumbum, and ferrum.

Some common elements and their symbols:

Hydrogen H
Oxygen O
Nitrogen N
Sulfur S
Carbon C
Calcium Ca
Chlorine Cl
Cobalt Co
Copper Cu (from the name cuprum)
Lead Pb (from the name plumbum)

Table 16-1 is an alphabetical list of the elements.

Atom

The smallest particle of an element is known as an atom. An atom is a very small particle, being formed of a central core, called *nucleus,* surrounded by a cloud of revolving tiny particles, the *electrons.*

The nucleus consists mainly of two types of particles: positively charged *protons* and neutral *neutrons.* An electron carries a negative charge and is insignificant in weight. In an uncombined atom, the number of protons equals the number of electrons. Thus, an atom is neutral (positive charge equals the negative charge). Sets of electrons revolve around the nucleus at different distances, known as *shells.* The maximum number of shells is 7. Each shell can hold a certain maximum number of electrons. For example, the maximum number of electrons held by the first, second, and third shells, are 2, 8, and 18, respectively.

Each element has a specific number of protons in its atom, called the *atomic number.* For example, a hydrogen atom has 1 proton, helium has 2, carbon has 6, chlorine has 17, calcium has 20, and uranium has 92 protons in their respective nuclei. An element is identified by its atomic number.

A proton weighs about the same as a neutron. The sum of protons and neutrons in an atom is known as the *mass number* because this number determines the weight of an atom, called the *atomic weight.* The mass number is close to twice the atomic number and is approximately equal to the atomic weight. Each element has atoms with different number of neutrons (e.g., hydrogen atoms have 0, 1, or 2 neutrons). Atoms of the same element with different number of neutrons are known as its *isotopes.* An isotope is iden-

Table 16-1

Name	Symbol	Atomic Number	Atomic Weight	Electro- negativity
Actinium	Ac	89	227*	1.1
Aluminum	Al	13	26.98	1.5
Americium	Am	95	243*	1.3
Antimony	Sb	51	121.75	1.9
Argon	Ar	18	39.95	—
Arsenic	As	33	74.92	2.0
Astatine	At	85	210*	2.2
Barium	Ba	56	137.34	0.9
Berkelium	Bk	97	247*	1.3
Beryllium	Be	4	9.01	1.5
Bismuth	Bi	83	208.98	1.9
Bohrium	Bh	107	262	—
Boron	B	5	10.81	2.0
Bromine	Br	35	79.90	2.8
Cadmium	Cd	48	112.40	1.7
Calcium	Ca	20	40.08	1.0
Californium	Cf	98	249*	1.3
Carbon	C	6	12.01	2.5
Cerium	Ce	58	140.12	1.1
Cesium	Cs	55	132.91	0.7
Chlorine	Cl	17	35.45	3.0
Chromium	Cr	24	52.00	1.6
Cobalt	Co	27	58.93	1.8
Copper	Cu	29	63.55	1.9
Curium	Cm	96	247*	1.3
Dubnium	Db	105	262	—
Dysprosium	Dy	66	162.50	1.1
Einsteinium	Es	99	254*	1.3
Erbium	Er	68	167.26	1.1
Europium	Eu	63	151.96	1.1
Fermium	Fm	100	253*	1.3
Fluorine	F	9	19.00	4.0
Francium	Fr	87	223*	0.7
Gadolinium	Gd	64	157.25	1.1
Gallium	Ga	31	69.72	1.6
Germanium	Ge	32	72.59	1.8
Gold	Au	79	196.97	2.4
Hafnium	Hf	72	178.49	1.3
Hassium	Hs	108	265	—
Helium	He	2	4.00	—
Holmium	Ho	67	164.93	1.1
Hydrogen	H	1	1.01	2.1
Indium	In	49	114.82	1.7
Iodine	I	53	126.90	2.5
Iridium	Ir	77	192.22	2.2
Iron	Fe	26	55.85	1.8
Krypton	Kr	36	83.80	—
Lanthanum	La	57	138.91	1.1
Lawrencium	Lr	103	257*	—
Lead	Pb	82	207.2	1.8
Lithium	Li	3	6.94	1.0
Lutetium	Lu	71	174.97	1.2
Magnesium	Mg	12	24.31	1.2
Manganese	Mn	25	54.94	1.5
Meitnerium	Mt	109	265	—
Mendelevium	Md	101	256*	—

* Mass number of most stable or best-known isotope.
† Mass of most commonly available, long-lived isotope.

Table 16-1 (*Continued*)

Name	Symbol	Atomic Number	Atomic Weight	Electro- negativity
Mercury	Hg	80	200.59	1.9
Molybdenum	Mo	42	95.94	1.8
Neodymium	Nd	60	144.24	1.1
Neon	Ne	10	20.18	—
Neptunium	Np	93	237.05†	1.3
Nickel	Ni	28	58.71	1.8
Niobium	Nb	41	92.91	1.6
Nitrogen	N	7	14.01	3.0
Nobelium	No	102	254*	1.3
Osmium	Os	76	190.2	2.2
Oxygen	O	8	16.00	3.5
Palladium	Pd	46	106.4	2.2
Phosphorus	P	15	30.97	2.1
Platinum	Pt	78	195.09	2.2
Plutonium	Pu	94	242*	1.3
Polonium	Po	84	210*	2.0
Potassium	K	19	39.10	0.8
Praseodymium	Pr	59	140.91	1.1
Promethium	Pm	61	147*	1.1
Protactinium	Pa	91	231.04*	1.5
Radium	Ra	88	226.03†	0.9
Radon	Rn	86	222*	—
Rhenium	Re	75	186.2	1.9
Rhodium	Rh	45	102.91	2.2
Rubidium	Rb	37	85.47	0.8
Ruthenium	Ru	44	101.07	2.2
Rutherfordium	Rf	104	261	—
Samarium	Sm	62	150.4	1.1
Scandium	Sc	21	44.969	1.3
Seaborgium	Sg	106	263	—
Selenium	Se	34	78.96	2.4
Silicon	Si	14	28.09	1.8
Silver	Ag	47	107.87	1.9
Sodium	Na	11	22.99	0.9
Strontium	Sr	38	87.62	1.0
Sulfur	S	16	32.06	2.5
Tantalum	Ta	73	180.95	1.5
Technetium	Tc	43	98.91†	1.9
Tellurium	Te	52	127.60	2.1
Terbium	Tb	65	158.93	1.1
Thallium	Tl	81	204.37	1.8
Thorium	Th	90	232.04†	1.3
Thulium	Tm	69	168.93	1.1
Tin	Sn	50	118.69	1.8
Titanium	Ti	22	47.90	1.5
Tungsten	W	74	183.85	1.7
Uranium	U	92	238.03	1.7
Vanadium	V	23	50.94	1.6
Xenon	Xe	54	131.30	—
Ytterbium	Yb	70	173.04	1.1
Yttrium	Y	39	88.91	1.2
Zinc	Zn	30	65.38	1.6
Zirconium	Zr	40	91.22	1.4

* Mass number of most stable or best-known isotope.
† Mass of most commonly available, long-lived isotope.

tified by writing the mass number after the name of the element (e.g., oxygen-16, and carbon-12). Atomic weight of an element is the average atomic weight of its isotopes.

Atoms are very small and light, so they are weighed in very small units, the atomic mass units (amu). One amu equals one twelfth of the weight of one carbon-12 atom, which is assigned the weight 12 amu. Carbon-12 is the standard for determining the atomic weights of elements. Atomic mass unit is used for weighing atoms just as a gram is used for weighing precious metals and a pound is used for weighing groceries. For practical purposes, elements are weighed in grams. When gram replaces atomic mass unit, the number of the atoms changes from one to the number of amu in 1 gram.

$$1 \text{ gram} = 602{,}000{,}000{,}000{,}000{,}000{,}000{,}000 \text{ amu}$$
$$= 6.02 \times 10^{23} \text{ amu, (in scientific notation form).}$$

This constant is known as *Avogadro number*. Gram atomic weight of an element is called its *mole*. For example, moles of carbon-12 and hydrogen-1 are 12 grams and 1 gram, respectively.

Periodic Table

The periodic table is the table of elements arranged according to their atomic numbers (number of protons in the nucleus). Table 16-2 shows the periodic table. It starts with hydrogen with the atomic number 1 and ends with the element with atomic number 112 (the highest known atomic number). Each element is assigned a block, having information about the element. The vertical columns of elements are known as *groups* or families, and the horizontal rows are called *periods* or series. Groups are identified by roman numerals such as IA-VIIIA, indicating the number of electrons in the outermost shell, so elements of each group have the same number of electrons in their outermost shell. The properties of an element are determined by the electrons in its outermost shell. Thus, elements in the same group have similar properties, just as members of the same family have similar traits.

Periodic Law

Chemical properties of elements are repeated periodically as the atomic numbers increase. For example, sodium with atomic number 11 and potassium with atomic number 19 have similar properties. They belong to the same group IA. Each group starts with an element with one shell and ends with an element having seven shells.

Each period has the same number of shells. Periods are identified by the Arabic numerals such as 1, 2, 3, and 7, indicating the number of the outermost shell. Period 1 has one shell (with only two elements, hydrogen and helium), period 2 has two shells, and period 7 has seven shells. A period starts with

Table 16-2

IA	IIA	IIIB	IVB	VB	VIB	VIIB	VIII	VIII	VIII	IB	IIB	IIIA	IVA	VA	VIA	VIIA	Gases
1 **H** 1.00797																1 **H** 1.00797	2 **He** 4.0026
3 **Li** 6.939	4 **Be** 9.0122											5 **B** 10.811	6 **C** 12.0112	7 **N** 14.0067	8 **O** 15.9994	9 **F** 18.9984	10 **Ne** 20.183
11 **Na** 22.9898	12 **Mg** 24.312											13 **Al** 26.9815	14 **Si** 28.086	15 **P** 30.9738	16 **S** 32.064	17 **Cl** 35.453	18 **Ar** 39.948
19 **K** 39.102	20 **Ca** 40.08	21 **Sc** 44.956	22 **Ti** 47.90	23 **V** 50.942	24 **Cr** 51.996	25 **Mn** 54.9380	26 **Fe** 55.847	27 **Co** 58.9332	28 **Ni** 58.71	29 **Cu** 63.54	30 **Zn** 65.37	31 **Ga** 69.72	32 **Ge** 72.59	33 **As** 74.9216	34 **Se** 78.96	35 **Br** 79.909	36 **Kr** 83.80
37 **Rb** 85.47	38 **Sr** 87.62	39 **Y** 88.905	40 **Zr** 91.22	41 **Nb** 92.906	42 **Mo** 95.94	43 **Tc** (99)	44 **Ru** 101.07	45 **Rh** 102.905	46 **Pd** 106.4	47 **Ag** 107.870	48 **Cd** 112.40	49 **In** 114.82	50 **Sn** 118.69	51 **Sb** 121.75	52 **Te** 127.60	53 **I** 126.904	54 **Xe** 131.30
55 **Cs** 132.905	56 **Ba** 137.34	*57 **La** 138.91	72 **Hf** 178.49	73 **Ta** 180.948	74 **W** 183.85	75 **Re** 186.2	76 **Os** 190.2	77 **Ir** 192.2	78 **Pt** 195.09	79 **Au** 196.967	80 **Hg** 200.59	81 **Tl** 204.37	82 **Pb** 207.19	83 **Bi** 208.980	84 **Po** (210)	85 **At** (210)	86 **Rn** (222)
87 **Fr** (223)	88 **Ra** (226)	‡89 **Ac** (227)	104 **Rf** (261)	105 **Db** (262)	106 **Sg** (266)	107 **Bh** (262)	108 **Hs** (265)	109 **Mt** (266)	110 **?** (271)	111 **?** (272)	112 **?** (277)						

* Lanthanide Series

58 **Ce** 140.12	59 **Pr** 140.907	60 **Nd** 144.24	61 **Pm** (147)	62 **Sm** 150.35	63 **Eu** 151.96	64 **Gd** 157.25	65 **Tb** 158.924	66 **Dy** 162.50	67 **Ho** 164.930	68 **Er** 167.26	69 **Tm** 168.934	70 **Yb** 173.04	71 **Lu** 174.97

‡ Actinide Series

90 **Th** 232.038	91 **Pa** (231)	92 **U** 238.03	93 **Np** (237)	94 **Pu** (242)	95 **Am** (243)	96 **Cm** (247)	97 **Bk** (247)	98 **Cf** (249)	99 **Es** (254)	100 **Fm** (253)	101 **Md** (256)	102 **No** (256)	103 **Lr** (257)

Numbers in parenthesis are mass numbers of most stable or most common isotope.

Atomic weights corrected to conform to the 1963 values of the Commission on Atomic Weights.

The group designations used here are the former Chemical Abstract Service numbers.

an element having one electron in its outermost shell and ends with an element having two or eight electrons, which are called the *stable state*. An unstable outermost shell is known as *valence shell* and its electrons as *valence electrons*.

The maximum number of shells is 7 and the maximum number of electrons in the outermost shell is 8; there are, therefore, seven periods and eight groups.

The periodic table is a useful tool to understand the properties of the elements, such as atomic number, atomic weight, and whether the element is a metal or a nonmetal. Elements with one to three valence electrons are known as metals (except hydrogen) and those with five to seven as nonmetals. Elements of group VIII (last group) are stable (have two or eight electrons in the outermost shell) and are known as inert or noble gases because they do not react with other elements.

COMPOUNDS

Elements form compounds to be chemically stable, by having a stable outermost shell (with two or eight electrons). A compound is formed of two or more elements chemically combined in a definite proportion, which is called the *law of definite composition*. The law states that a compound has a definite composition by weight. For example, carbon dioxide always and everywhere is formed of 27.3 percent carbon and 72.7 percent oxygen, and water is always 11.1 percent hydrogen and 88.9 percent oxygen. Some examples of compounds are water, lime, alum, and sodium hypochlorite. Compounds are formed by transfer or sharing of valence electrons between their atoms.

Valence or Oxidation Number

The valence or oxidation number is the number of electrons being shared or transferred by an atom in the compound formation. The transfer of electrons forms ionic compounds, and sharing of electrons forms covalent compounds.

Ionic Compounds. In the sodium chloride formation, sodium has only one valence electron and chlorine has seven valence electrons (see Figure 16-1). In this reaction, the valence electron of sodium transfers to valence shell of the chlorine atom. Thus, the sodium atom becomes stable by losing its single valence electron in the third shell (as the second shell is stable with eight electrons) to chlorine atom, and chlorine atom becomes stable by gaining that electron.

When electrons are transferred, atoms become charged. The charged particles are known as *ions*. The positive ion is known as a *cation* and the negative ion as an *anion*. Thus, sodium atom becomes a *cation* with one positive charge by losing its electron, and chlorine atom becomes an anion

Table 16-3   Ions and Their Oxidation Numbers

Cations

Oxidation Number

+1		+2		+3	
Ammonium	NH_4^+	Barium	Ba^{++}	Aluminum	Al^{+++}
Lithium	Li^+	Calcium	Ca^{++}	Iron (III), ferric	Fe^{+++}
Sodium	Na^+	Copper (II), cupric	Cu^{++}		
Potassium	K^+	Iron (II), ferrous	Fe^{++}		
Silver	Ag^+	Lead	Pb^{++}		
Copper (I), cuprous	Cu^+	Magnesium	Mg^{++}		
		Nickel (II)	Ni^{++}		
		Zinc	Zn^{++}		

Anions

Oxidation Number

−1		−2		−3	
Acetate	$C_2H_3O_2^-$	Carbonate	$CO_3^=$	Phosphate	$PO_4^{\equiv}$
Azide	N_3^-	Chromate	$CrO_4^=$		
Chloride	Cl^-	Dichromate	$Cr_2O_7^=$		
Bromide	Br^-	Oxide	$O^=$		
Iodide	I^-	Peroxide	$O_2^=$		
Hydroxide	OH^-	Sulfate	$SO_4^=$		
Hydrogen Carbonate (Bicarbonate)	HCO_3^-	Sulfide	$S^=$		
Hydrogen Sulfate (Bisulfate)	HSO_4^-	Sulfite	$SO_3^=$		
Nitrate	NO_3^-	Thiosulfate	$S_2O_3^=$		
Nitrite	NO_2^-				
Chlorate	ClO_3^-				
Hypochlorite	ClO^- or OCl^-				

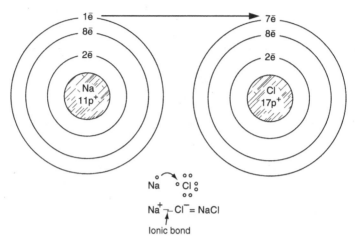

Figure 16-1 Sodium Chloride Formation

(*Source: Basic Chemistry for Water and Wastewater Operators.* Copyright © 2002, American Water Works Association. Reprinted by permission.)

with one negative charge by gaining that electron. The gaining of an electron is called *reduction,* and the losing of an electron is *oxidation.* A substance that loses electrons is called a *reducing agent,* and one that gains electrons is known as an *oxidizing agent.*

In the sodium chloride formation, it is oxidation of sodium and reduction of chlorine. Sodium is the reducing agent, and chlorine is the oxidizing agent. Metals are reducing agents because they lose electrons, and nonmetals are oxidizing agents as they gain electrons. Examples of ionic compounds are sodium chloride, calcium oxide, and magnesium chloride.

Covalent Compounds. When nonmetals combine with nonmetals, electrons are shared, and covalent bonds are formed. A pair of shared electrons forms a covalent bond. Examples of covalent compounds are water, carbon dioxide, ammonia, and organic compounds.

Organic compounds are covalent compounds with carbon as a common element. They are commonly present in plants and animals. Most of them are formed of carbon, hydrogen, oxygen, and nitrogen. Common examples are starches, sugars, proteins, fats, and oils.

Carbon has four covalent bonds, nitrogen has three, oxygen has two, and hydrogen has one. Organic compounds often have long chains or rings of carbon bonded with O, H, and N. Thus, they have a high molecular weight. As shown in Figure 16-2, they are represented by structural formulae (showing actual bonds between atoms).

Chemical Formula

A chemical formula is a symbolized representation of the chemical composition of a compound, by using chemical symbols and numerical subscripts.

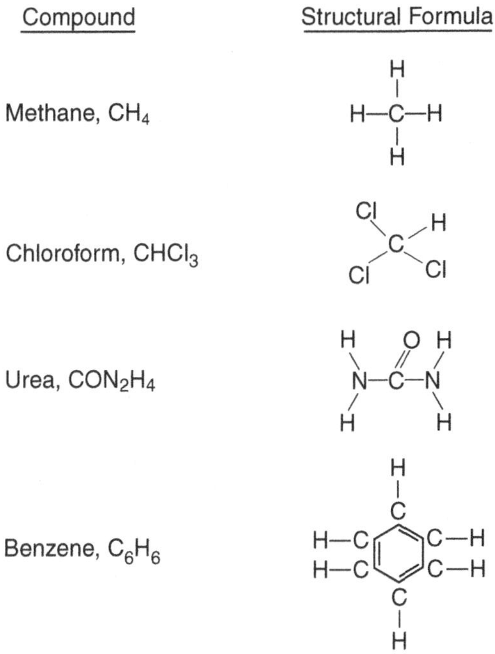

Compound	Structural Formula

Methane, CH_4

Chloroform, $CHCl_3$

Urea, CON_2H_4

Benzene, C_6H_6

Figure 16-2 Organic Compounds

A formula has two parts: the positive and the negative. The positive part is written first and the negative part second. Positive charge on the first part is equal to the negative charge on the second part, which is achieved by applying appropriate subscripts if needed. Net charge on a formula is 0.

A *radical* is a charged group of combined atoms. Radicals are mostly anions. Common examples of radicals are carbonates, bicarbonates, sulfates, and phosphates. When a radical requires a subscript, it is parenthesized and the subscript is applied outside the parenthesis (see the formula for aluminum sulfate).

Examples

Compound		Formula
Sodium chloride	$Na^{+1}\ Cl^{-1}$	$NaCl$
Calcium chloride	$Ca^{+2}\ Cl^{-1}$	$CaCl_2$
Calcium carbonate	$Ca^{+2}\ CO_3^{-2}$	$CaCO_3$
Aluminum sulfate	$Al^{+3}\ SO_4^{-2}$	$Al_2(SO_4)_3$

We can write different formulae by using the cation and anion table, Table 16-3.

Naming Compounds

In binary compounds (formed of two elements), a name is formed using the full name of the first element and the root of the name of the second element ending with -ide. Examples of binary compounds are sodium chloride, calcium oxide, calcium chloride, and hydrogen sulfide. When there are three or more elements forming a compound (normally the second part is a radical), then the complete names of the both parts of formula are used, such as aluminum sulfate and calcium carbonate.

Percentage Composition

Percentage composition of a compound is the percentage weight of each element in the formula of a compound. It is the total atomic weight of each element divided by the formula weight (total weight of all the atoms in the formula) of the compound, and then multiplied with 100.

Formula Weight	Percentage Composition
Water, H_2O $(2 \times 1) + 16 = 18$	Hydrogen $= (2/18) \times 100 = 11.11\%$
	Oxygen $= (16/18) \times 100 = 88.89\%$
Carbon dioxide, CO_2 $12 + (2 \times 16) = 44$	Carbon $= (12/44) \times 100 = 27.27\%$
	Oxygen $= (32/44) \times 100 = 72 \cdot 73\%$

Gram formula weight of a compound is its mole. For example, moles of water and carbon dioxide are 18 grams and 44 grams, respectively.

CHEMICAL EQUATIONS

A chemical equation is a concise and symbolized representation of a chemical reaction. Chemicals which react together are called *reactants,* and those which are produced are known as *products.* Reactants are written on the left side and products on the right side of an arrow, showing the direction of the reaction.

Example

$$\text{Sodium} + \text{Chlorine} \rightarrow \text{Sodium chloride}$$

$$2Na \quad + \quad Cl_2 \quad \rightarrow \quad 2\,Na\,Cl$$
$$2 \times 23 \quad 2 \times 35.5 \quad 2 \times (23 + 35.5)$$

Mass ratio: 46   :   71   :   117

Significance of the Arrows

$\rightarrow$ = reaction proceeds to form products
$\leftrightarrows$ = reaction is reversible, meaning products revert back to reactants
$\downarrow$ = product precipitates out
$\uparrow$ = product is a gas

A chemical reaction follows the law of conservation of masses. The law states: *In a chemical reaction, the total mass of reactants equals the total mass of the products.* These masses come from various atoms; thus, it is the conservation of each type of atoms on both sides of the equation. In a chemical reaction, the atoms of reactants are rearranged to form various products.

To satisfy the law, appropriate coefficients are assigned (in small whole number ratio) to each chemical in the equation to balance it by making the number of each type of atoms the same on both sides. In the previous equation, coefficient 2 is needed in front of Na and Na Cl to balance the equation.

Hydrogen, oxygen, nitrogen, fluorine, chlorine, bromine, and iodine elements exist in nature as diatomic molecules, formed by sharing their own electrons (H_2, O_2, N_2, F_2, Cl_2, Br_2, and I_2). A molecule is the smallest stable and neutral particle of a substance. In chemical reactions, these elements are represented as diatomic molecules. In the given equation, chlorine reactant has one molecule or two atoms as the minimum to participate in the reaction.

Significance of a Chemical Equation

A chemical equation shows both the quality and quantity of the reactants and their products. In the previous example, two sodium atoms and one molecule of chlorine react to form two formulas of sodium chloride. By assigning appropriate masses, we learned that 46, (2×23), mass units (amu, grams, or pounds) of sodium react with 71, (2×35.5), mass units of chlorine to produce 117, (2×78.5), mass units of sodium chloride. This quantitative relationship is known as *stoichiometery,* which is the calculation of relative weights of reactants and their products in a chemical reaction. Stoichiometery is useful for chemical dose calculations.

A reaction is not possible if the equation cannot be balanced. From a balanced equation, we can calculate the mass of any reactant or a product corresponding to a known mass of one of them. The mass ratio of reactants and their products remains the same.

In the previous reaction, if we use only 35.5 grams of chlorine, amounts of sodium and sodium chloride will be 23 and 58.5 grams, respectably.

$$X \qquad 35.5\ g \qquad X_1$$
$$2Na\ +\quad Cl_2 \quad \rightarrow 2NaCl$$
$$46\ :\quad 71\quad :\quad 117$$

$$X = (35.5 \text{ g}/71 \text{ g}) \times 46 \text{ g} = 23 \text{ g}$$
$$X_1 = (35.5 \text{ g}/71 \text{ g}) \times 117 \text{ g} = 58.5 \text{ g}$$

where:

X = the new mass of sodium
X_1 = the new mass of sodium chloride

Example. How many mg/L of CO_2 will react with 50 mg/L of lime, $Ca(OH)_2$?

$$\underset{40}{CO_2} + \underset{74}{Ca(OH)_2} \rightarrow \underset{100}{Ca\,CO_3\downarrow} + \underset{18}{H_2O}$$

with X above CO_2 and 50 mg/L above $Ca(OH)_2$

$$X/40 \text{ mg/L} = 50 \text{ mg/L}/74 \text{ mg/L}$$
$$X = (50 \text{ mg/L}/74 \text{ mg/L}) \times 40 \text{ mg/L} = 29.7 \text{ mg/L of } CO_2$$

where:

X = new mass of CO_2

ACIDS, BASES, AND SALTS

Acids

Acids are covalent compounds, with hydrogen as the first part of the formula. Their names end with -ic or -ous. Examples are hydrochloric acid (HCl), sulfuric acid (H_2SO_4), nitric acid (HNO_3), and hypochlorous (HOCl) acid. They form hydrogen ions (H^+), the acidity, in their water solutions. Ion formation by acids is known as *ionization*. As a hydrogen ion is just the nucleus of the hydrogen atom with one proton, so the hydrogen ion is called here a proton. Therefore, acids are proton donors. Generally, the more the number of hydrogen atoms in the formula, the less the ionization and the weaker is the acid, and vice versa. Hydrochloric acid readily ionizes in water and is a strong acid; phosphoric acid (H_3PO_4) ionizes poorly and is a weak acid.

Bases

Bases are chemicals that accept protons (H^+) from acids. Bases neutralize acids. Common examples of bases are sodium hydroxide, calcium hydroxide, and ammonium hydroxide.

Salts

Salts are formed when acids react with bases. A salt is an ionic compound. Some examples of salts are chlorides, bromides, iodides, sulfates, and phosphates. A salt is formed of the first part of the formula of a base and second part of the formula of an acid.

Example. Hydrochloric acid reacts with sodium hydroxide to produce the salt, sodium chloride, and water. In this reaction, H^+ of the acid combines with OH^- of the base to form water, and Na^+ and Cl^- ions form sodium chloride (NaCl).

$$\text{Acid} \quad \text{Base} \qquad \text{Salt}$$
$$\text{HCl} + \text{NaOH} \rightarrow \text{NaCl} + \text{H}_2\text{O}$$

pH

pH is the minus log (negative power of the base 10) of the hydrogen ion concentration in a solution as moles/liter.

$pH = -\log [H^+]$, the brackets [ ] stand for concentration as mol/L.

If an acid has 1/10 mol/L of H^+, its hydrogen ion concentration is 10^{-1} mol/L. So, power of the base 10 is −1, and pH is 1. If the H^+ concentration is 1/100 or 10^{-2} mol/L, then pH is 2. See Figure 16-3 for a pH scale.

The pH scale is based on the self-ionization of water to form H^+ and OH^- ions. This concentration is 10^{-7} mol/L for each of these ions at room temperature. Product of their concentration is 10^{-14}. Thus, pH scale ranges from 0 to 14. At pH 0, H^+ concentration is 10^0 or 1 mol/L, and at pH 14 it is 10^{-14} mol/L; OH^- ion concentration at these pH values is 10^{-14} mol/L and 10^0 mol/L, respectively. The pH scale measures acidity or basidity up to only 1 mol/L of their respective ions. At pH 7, H^+ concentration equals OH^- concentration; thus, the pH is neutral. Acidity increases from pH 7 to 0, and basidity increases from pH 7 to 14. Each unit decrease in pH equals 10 times more acidity, and vice versa. The lower the pH, the higher is the acidity.

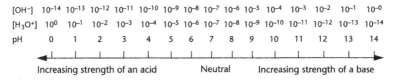

Figure 16-3 pH Scale

(*Source: Basic Chemistry for Water and Wastewater Operators*, Copyright © 2002, American Water Works Association. Reprinted by permission.)

$$H_2O \leftrightharpoons H^+ + OH^-$$

$$[10^{-7} \text{ mol/L}] \, [10^{-7} \text{ mol/L}] = 10^{-14}$$

SOLUTIONS

A solution is a homogeneous (uniform concentration) mixture of two or more substances. The dissolved substance is known as a *solute* and the dissolving medium as a *solvent*. For example, sodium chloride dissolved in water is a solute, and the water is a solvent. A solution with the maximum concentration of its solute is known as a *saturated* solution.

Concentration of a Solution

Percent concentration is parts of the solute in 100 parts of the solution. For example, 10 grams of a salt in 100 grams of solution is a 10 percent salt solution.

Molarity (M) is the number of moles of the solute/L of the solution, e.g., 0.01 mol/L (0.98 g/L) of sulfuric acid is 0.01 molar sulfuric acid solution. It is written as 0.01M H_2SO_4.

Normality (N) of a solution is the number of gram equivalent weights of the solute/L of the solution. *Gram-equivalent weight* of a substance is its weight in grams, which contributes one Avogadro number of electrons in a chemical reaction. Thus, these weights are chemically equivalent. Generally, gram-equivalent weight of a substance is its mole divided by the positive charge on the first part of the formula.

Substance	Mole	Gram-Equivalent Weight
HCl	36.5 g	36.5/1 = 36.5 g
H_2SO_4	98 g	98/2 = 49 g
H_3PO_4	98 g	98/3 = 32.6 g
$CaCO_3$	100 g	100/2 = 50 g
NaOH	40 g	40/1 = 40 g

Standard Solutions

A standard solution has the precise concentration of the solute. It is used, in the laboratory, to determine the concentration of other solutions by the titration technique. A standard solution is called a *titrant,* and the technique to match it with a solution of an unknown concentration is known as *titration.* The technique is based on the fact that e*qual volumes of solutions with the same normality are chemically equivalent.* In titration, normality multiplied by the volume of the standard solution equals the normality multiplied by the volume of the standardized solution.

$$N \times V = N_1 \times V_1$$

where:

N = normality, corresponding to the volume V of the standard
N_1 = normality, corresponding to volume V_1 of the standardized

The equivalence or end point is determined by an indicator, a chemical that changes color in standard and standardized solutions. For example, in the alkalinity test, phenolphthalein is pink in alkaline and colorless in acidic solutions.

Example. Forty mL of NaOH solution is titrated with 20 mL of 0.02 N H_2SO_4, standard solution. Determine the normality of the NaOH solution.

$$N \times 40 \text{ mL} = 20 \text{ mL} \times 0.02\ N$$

Normality of NaOH = (20 mL $\times$ 0.02 N)/40 mL = 0.01N

where:

N = normality of NaOH

This principle is also used in the laboratory to make dilute solutions from the stock solutions.

QUESTIONS

1. Define the terms *element, compound, atom, proton, neutron, electron, atomic number,* and *atomic weight.*

2. Write symbols and atomic weights for the following elements:
   a. Sodium
   b. Potassium
   c. Magnesium
   d. Iron
   e. Copper
   f. Lead
   g. Chlorine
   h. Gold

3. Explain the terms: *valence electrons, oxidation number, group, period,* and *inert gases.*

4. Write the differences between the following terms:

    a. Organic and inorganic compounds

    b. Oxidation and reduction

    c. Oxidizing and reducing agents

5. Determine the chemical composition of aluminum sulfate.

6. Explain the meaning of these arrows in the chemical equations.

    a. →

    b. ⇋

    c. ↑

7. a. Define the terms *acid, base, salt,* and *pH.*

8. The lower the pH, the higher is the acidity. T or F

9. In a titration, it took 10 mL of 1*N* standard solution of hydrochloric acid to neutralize 40 mL of sodium hydroxide solution. Determine the normality of the sodium hydroxide solution.

10. Differentiate between molarity and normality of a solution. What is the normality of a 0.01M solution of sulfuric acid?

17

MICROBIOLOGY

Microbiology is the science of microscopic plants and animals called *microorganisms* or *microbes*. An organism, living thing, has characteristic shape and size, metabolism, reproduction, movements, growth, and adaptation as its important characteristics, which separate it from nonliving things. Metabolism consists of various chemical changes that maintain the life activities of an organism. All waterborne pathogens are microbes. The major concern in producing safe drinking water is removing or inactivating the waterborne pathogens.

Anthony von Leeuwenhook, a draper from Holland, was the first to see microorganisms in 1674 with a microscope he developed that had 200 times magnification. Microbiology developed into a major branch of biology in the second half of nineteenth century, mainly due to the discoveries of Louis Pasteur and Robert Koch. Pasteur made three important discoveries in the 1870s and 1880s in France: bacteria come from preexisting bacteria, pasteurization (named after him), and immunization (vaccination). For pasterization, he heated beverages, such as beer, wine, and grape juice for 30 minutes at 62°C, then chilled them quickly to about 10°C. This process killed food spoiling bacteria, maintained the original taste of the beverage, and kept the beverage for a long time without getting spoiled. For immunization, he attenuated (aged) cultures of bacteria of diseases like chicken cholera, rabies, tuberculosis, cholera, and tetanus, then injected them into healthy hosts, which then developed immunity to the disease without the disease symptoms.

Robert Koch, a German biologist, worked on *Anthrax* bacterium in 1878. To obtain the isolated colonies, he developed methods to raise pure cultures of bacteria on the semisolid media. He made four important discoveries:

1. A specific organism is associated with a specific disease.
2. It can be isolated and grown as a pure culture in a lab.

3. The lab culture will produce the disease when injected into a healthy host.

4. The organism can be isolated from an infected host.

BINOMIAL NOMENCLATURE SYSTEM

According to the international binomial nomenclature system, each organism is given a two-word Latin or Latinized scientific name. The first word is the genus and the second word is the species. Genus, a noun, is capitalized. The species, an adjective, is lowercase. Both parts are written separately and are either individually underlined or italicized. Man is named *Homo sapiens* meaning wise man (*homo* = man and *sapiens* = wise). Other examples are *Escherichia coli* (*Escherichia* found in colon, a fecal coliform bacterium) and *Giardia lamblia*.

CELL

A *cell* is a unit of protoplasm, the living matter. *Protoplasm* is the Greek word for the first formed substance.

Cell Theory

The cell theory was developed in Germany by two biologists—Mathias Jakob Schleiden, a botanist; and Theodor Schwann, a zoologist—in 1839. This theory can be summarized as follows: organisms (both plants and animals) are formed of cells; a cell is a unit of structure and function; and all cells come from preexisting cells. Before this, Robert Hooke was the first person to see cells in a piece of cork with a microscope, in 1665. The microscope was developed by Christopher Cook with 14–42 times magnification.

Cell Structure

A cell is formed of two main parts: nucleus and cytoplasm. Nucleus, the central part, is the most important part of the cell because it carries all the genetic code of the organism in the form of chromosomes. Chromosomes are formed of chemicals, known as deoxyribonucleic acids (DNAs). Each chromosome is formed as a set of *genes,* which are units of genetics. Each gene carries information about a specific trait of the organism. The nucleus is generally covered with a nuclear membrane and is surrounded by the cytoplasm. The cell is covered with a protective cytoplasmic membrane that is semipermeable, meaning that it allows water to pass in or out of the cell, and only selective chemicals can get through. All the chemical functions take place in the cytoplasm (see Figure 17-1). The plant cells differ from animal cells in having a rigid cell wall around the cell membrane.

Microbes, unlike higher plants and animals, do not have their bodies differentiated into different tissues (groups of similar cells with a specific function) and organs for different life functions. Microbes are unicellular; therefore, all life activities are carried out by the same cell.

Microbes consist of five main groups: viruses, bacteria, fungi, algae, and protozoa. Bacteria, being common waterborne pathogens and indicator organisms of fecal contamination, will be discussed in detail. The other groups will be covered only briefly.

VIRUSES

The word *virus* means poison or venom. Viruses are the smallest known organisms. They are the borderline case between living and nonliving things. Their size varies from 10 to 300 nanometers; therefore, they pass through filters and are hard to remove in the water treatment. Viruses are intracellular obligate (strictly) parasites of plants and animals, and most of them are not susceptible to antibiotics. They live only inside the host cell. Outside the host, they are inactive, like nonliving matter. Viruses are very simple in their structure and very complex chemically. A *virus* consists only of a central core of

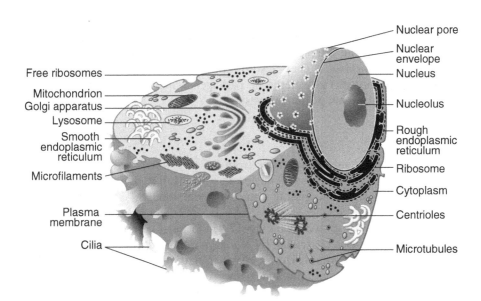

Figure 17-1 A Typical Cell

(*Source:* Alters & Alters, *Biology: Understanding Life.* Copyright © 2005 by John Wiley & Sons, Inc. Reprinted by permission of John Wiley & Sons, Inc.)

nucleic acids (deoxyribonucleic acids, DNA, or ribonucleic acids, RNA) covered with a sheath of proteins called *capsid*. A virus controls metabolism of the host cell, multiplies inside it, and then invades new cells. Thus, the virus converts the whole infected host cell into its own individuals (see Figure 17-2).

A viral infection can cause the host to develop immunity for the rest of its life. Therefore, the best preventive measure against a viral disease is vaccination. *Vaccination* is the injection of an attenuated (tamed or old) culture of a virus into the host to produce the immunity and not the disease symptoms.

Infectious hepatitis is the only confirmed viral waterborne disease. It is caused by the hepatitis A virus. Poliomyelitis is another example of a viral disease, which is suspected to be waterborne.

A high pH (above 10) and 0.5 to 1 mg/L free residual chlorine as hypochlorous acid are known to inactivate most viruses in the water. Lime softening and alum coagulation have been reported to remove 95 to 99 percent of viruses. Enteric viruses (viruses present in our large intestine) are regulated by the U.S. EPA to be removed or inactivated by 99.99 percent (i.e., 4 log removal or inactivation) in the drinking water.

BACTERIA

Bacteria are unicellular plants because they have a cell wall. However, they are without chlorophyll. They are heterotrophic (depend on synthesized food); most of them are saprophytic, living on sap of dead organic matter. They are commonly known as decomposers because they recycle dead organic matter by converting it into simpler substances like carbon dioxide, water, nitrates, sulfates, and phosphates. They absorb food from their environment. Bacteria form the majority of the waterborne pathogens, and some of them are used as indicator organisms of human wastes.

Bacterial Cell

A bacterial cell is small and varies from 0.5 to 5 micrometers in size. Most bacteria have a slimy capsule outside the cell wall (see Figure 17-3). The central nuclear region is without a nuclear membrane.

Reproduction. As shown in Figure 17-4, the most common method of reproduction in bacteria is asexual, by *binary fission* (split into two). A bacterial cell grows to its maximum size by taking in nutrients from the medium. The nuclear region grows and elongates; the cell splits into two daughter cells each with half of the nuclear matter and cytoplasm of the parent cell. Under optimum environmental conditions, it takes about 20 to 30 minutes for a cell to divide. This time period is known as the *generation time*. This is the time interval required for the population of an organism to double. Thus, a single

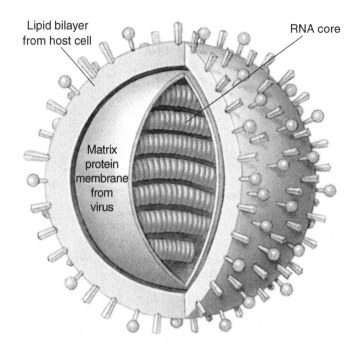

Lipid bilayer from host cell

RNA core

Matrix protein membrane from virus

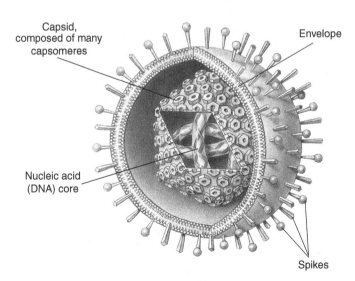

Capsid, composed of many capsomeres

Envelope

Nucleic acid (DNA) core

Spikes

Figure 17-2 Viruses

(*Source:* Black, *Microbiology, Sixth edition.* Copyright © 2005 by John Wiley & Sons, Inc. Reprinted by permission of John Wiley & Sons, Inc.)

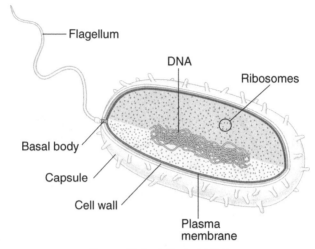

Figure 17-3 A Bacterial Cell

(*Source:* Alters & Alters, *Biology: Understanding Life.* Copyright © 2005 by John Wiley & Sons, Inc. Reprinted by permission of John Wiley & Sons, Inc.)

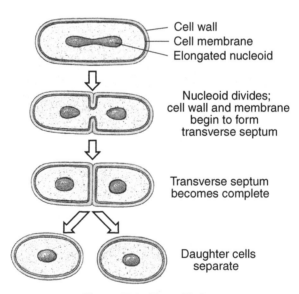

Figure 17-4 Binary Fission

(*Source:* Black, *Microbiology, Sixth edition.* Copyright © 2005 by John Wiley & Sons, Inc. Reprinted by permission of John Wiley & Sons, Inc.)

bacterium on a semisolid medium or filter forms a large spotlike colony within 24 hours, consisting of millions of its individuals. The number of bacteria in a water sample is determined by culturing them and counting their colonies.

Forms of Bacteria. All bacteria have one of the three forms: spherical, rod shaped, or spirallike (see Figure 17-5). They are called cocci, bacilli, and spirilli, respectively. Motility in bacteria is by flagella, whiplike extensions of cytoplasm. Flagella are present in all spirilli, in about half of the bacilli, and in only a few cocci.

Bacterial Growth in a Close System

In a closed system nothing comes in or goes out. To start with, a closed system has plenty of food, a few bacteria, and good conditions for their growth. Bacteria keep on growing until they reach the maximum number and biomass (their body weights). Then, gradually the growth decreases due to high death rate caused by the lack of food and buildup of harmful wastes. When plotted, the growth at different stages forms a specific curve.

The growth curve is formed of three phases, log growth, declining growth, and endogenous phase (see Figure 17-6). In the log growth phase, the food is plenty and the bacterium is growing and reproducing at the optimum rate. In the declining growth phase, the food supply is short, and there is buildup of harmful wastes, causing a progressive decrease in growth. In the endogenous phase, the growth stops; and bacteria starve and die, resulting in a decrease of biomass.

Classification of Bacteria Due to the Oxygen Requirement

- *Aerobic bacteria require oxygen.* The majority of bacteria, like higher organisms, utilize molecular oxygen for *respiration,* which is the oxidation (breakdown) of body substances by using oxygen to produce energy. Respiration produces carbon dioxide and water as its waste products. Respiration is comparable to burning of wood, where heat and light are produced along with carbon dioxide and water.
- *Anaerobic bacteria live without molecular oxygen.* They use energy from the fermentation (anaerobic decomposition of carbohydrates) or the putrefaction (anaerobic decomposition of proteins). They create foul smells due to nitrogen and sulfur containing byproducts, such as ammonia and hydrogen sulfide. Respiration is odorless and a more efficient energy producing system than the fermentation and putrefaction.
- *Facultative anaerobic bacteria can live in the presence or absence of molecular oxygen.* For their energy, they use respiration in the presence of oxygen and fermentation in its absence.

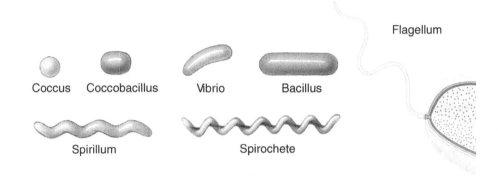

The most common bacterial shapes

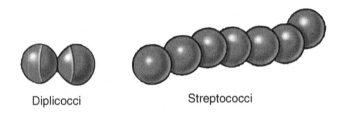

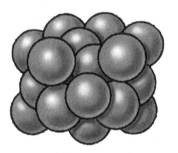

Staphylococci

Arrangements of bacteria

Figure 17-5 Forms of Bacteria

(*Source:* Black, *Microbiology, Sixth edition.* Copyright © 2005 by John Wiley & Sons, Inc. Reprinted by permission of John Wiley & Sons, Inc.)

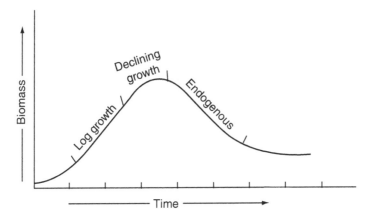

Figure 17-6 Bacterial Growth Curve

Bacterial Classes Due to the Temperature Preference

- *Psychrophilic bacteria prefer low temperature.* They can grow below 15°C. The optimum range of temperature is 15–20°C. They are, naturally, present in the colder regions.
- *Mesophilic bacteria grow at intermediate temperature.* Temperatures can be 15–45°C, with the optimum in the upper 30's. Most of the pathogenic and coliform bacteria belong to this group.
- *Thermophilic bacteria prefer high temperature such as above 45°C.* Their optimum temperature is in the 50's. They are common in tropical regions.

Means of Survival

Bacteria are sporulating and nonsporulating. Sporulating bacteria form *spores* under unfavorable conditions. Spores appear as small swellings either at the end or the middle of the cell. A spore is spherical or cylindrical structure containing dense protoplasm, mainly nuclear matter, surrounded by a tough layer of carbohydrates. Under the unfavorable conditions, it has a remarkable power of survival. When conditions are favorable, a single bacterial cell emerges from the spore and starts multiplying. Nonsporulating bacteria become dormant when conditions are unfavorable.

Coliform Bacteria

All bacteria present in the large intestine of man are known as *enteric bacteria,* which include pathogenic bacteria, some opportunistic or occasionally pathogenic, and a large number of saprophytic bacteria. Most of these bacteria are bacilli. The most common is *Escherichia coli* or *E. coli,* which is 0.5

micrometers wide and 1 to 3 micrometers long. *E. coli* and some other closely related enteric bacteria are known as *fecal coliform* bacteria. *E. coli* constitutes 80 to 95 percent of the fecal coliforms. Approximately, 2.0×10^{11} (200 billion) of *E. coli* are discharged daily in a person's fecal matter.

E. coli, being abundant, easy to identify, and relatively resistant of the waterborne pathogens, is used as an *indicator organism* (for the presence) of human wastes. *E. coli* is normally nonpathogenic; however, some strains can cause health problems. For example, in 1989, a strain of *E. coli* from a dairy cattle farm contaminated the drinking water of Cabool, Missouri, and caused 4 deaths and 240 cases of diarrhea. *E. coli* is also present in other warm-blooded animals, but it is not as abundant as in humans. Ordinarily, it dies off in the receiving waters after being discharged from the digestive system.

Another bacterium, very similar to *E. coli,* is *Enterobacter aerogenes* (*E. aerogenes*), which is found in soil contaminated with human wastes or sewage. *E. aerogenes* and some similar bacteria, coming from the contaminated environment, are known as *nonfecal coliforms*. Sometimes they are present in the biofilm of the water pipes. Coliform bacteria in the biofilm are predominantly *Klebsiella;* sometimes, they could be *Citrobacter* and *Enterobacter*. Presence of *E. coli* in the distribution system indicates either a breakdown in the treatment or a cross-connection because ordinarily *E. coli* cannot survive or grow there.

Fecal and nonfecal coliform bacteria together are called *total coliform group,* which is mainly formed of four genera: *Escherichia, Enterobacter, Klebsiella,* and *Citrobacter.*

Characteristics of Coliform Bacteria.   All coliform bacteria are aerobic and facultative anaerobic, nonsporulating, gram negative (stain red) bacilli, which ferment lactose (milk sugar) in 48 hours at 35°C.

Total coliform bacterial group is one of the primary drinking water standards. Water utilities have to test water for these bacteria routinely to determine its potability. According to the SDWA, the coliform test results should be negative in 95 percent of the water samples per month.

Bacterial Waterborne Pathogens

Following are the five known waterborne bacterial diseases and their pathogens:

Disease	Caused By
Cholera	*Vibrio cholerae*
Typhoid	*Salmonella typhosa*
Paratyphoid	*Salmonella paratyphi*
Bacillary dysentery	*Shigella dysenteriae*
Legionnaire's disease	*Legionella pneumophila*

Nuisance Bacteria

Mainly, nuisance bacteria are iron bacteria, sulfur bacteria, and actinomycetes. They cause tastes, odors, and corrosion in the distribution system. *Iron bacteria* are chemoautotrophic (see Figure 17-7). They utilize chemical energy for their food synthesis, by oxidizing soluble iron compounds in the water. They convert ferrous hydroxide to ferric hydroxide, the rust. Ferric hydroxide deposits in and on the bacterial slimy secretion, which gives a brownish color and unpleasant odor to the water. Common examples are *Crenothrix* and *Sphaerotilus.*

Sulfur bacteria use sulfur compounds for their metabolism. Two groups of sulfur bacteria—sulfate reducing and sulfide oxidizing—are important for a water utility. For example, *Desulfo vibrio* converts sulfates to hydrogen sulfide, which causes serious rotten-egg odors and corrosion, especially at dead ends and the stagnant water in the plumbing lines. Sulfide oxidizing bacteria convert hydrogen sulfide to sulfates, as in the case of aeration of water to remove hydrogen sulfide odors.

Examples of sulfur bacteria are *Chromatium* and *Thiospirillum.* These bacteria have sulfur granules in their cells (see Figure 17-8).

Actinomycetes are branched bacteria (bacterial cells are grouped to form branched structure) that decompose organic matter and produce *geosmin* and *methyl-iso barneol* (MIB). These chemicals cause earthy-musty odors in the surface waters, which are easily noticeable and hard to remove.

FUNGI

Fungi are multicellular lower plants (body is not differentiated into roots, stem, and leaves) without chlorophyll. Like bacteria, they are decomposers. They prefer low pH and a humid, dark environment. They grow on starchy foods. Fungi are either saprophytes or parasites (depend on another organism

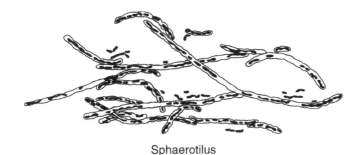

Sphaerotilus

Figure 17-7 Iron Bacteria

(*Source: Standard Methods for the Examination of Water and Wastewater, 20th edition.* American Public Health Association, 1999.)

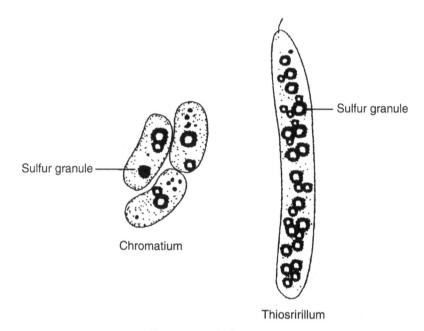

Figure 17-8 Sulfur Bacteria

(*Source: Standard Methods for the Examination of Water and Wastewater, 20th edition.* American Public Health Association, 1999.)

for food). Their body is formed of threadlike tubular structures known as *hyphae,* which are interwoven to form a network called a *mycelium.* They decompose plant matter to form humus. Sometimes, they grow in water lines and basins, causing tastes and odors. Common examples of fungi are yeast, molds, mildews, smuts, rusts, and mushrooms. Normally, fungi are not a problem in the water utility. Proper housekeeping and proper level of disinfectant keep them under control (see Figure 17-9).

ALGAE

Alga is the Latin word for seaweed. Algae are unicellular and multicellular lower aquatic plants, with chlorophyll. Like higher plants, they synthesize their food by photosynthesis. During daytime, their photosynthetic activity causes higher pH and higher dissolved oxygen in water. These conditions are unfavorable for some pathogenic bacteria. They also remove some hardness from the water.

They grow in lakes, reservoirs, rivers, and streams, causing color, tastes, and odors. During summertime, rapid multiplication of algae, called *algal blooms,* can cause serious taste and odor problems. Blue-green algae are the worst to cause this problem. Furthermore, some algae cause filter clogging.

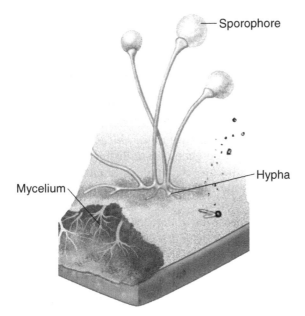

Figure 17-9 A Fungus

(*Source:* Alters & Alters, *Biology: Understanding Life,* Copyright © 2005 by John Wiley & Sons, Inc. Reprinted by permission of John Wiley & Sons, Inc.)

Classification of Algae Based on the Dominant Color

- *Cyanophyta* (*blue-green algae*) are unicellular or multicellular blue-green algae. See Figure 17-10. Chlorophyll is diffused in the cytoplasm. Like bacteria, cyanophyta do not have a nuclear membrane; they reproduce

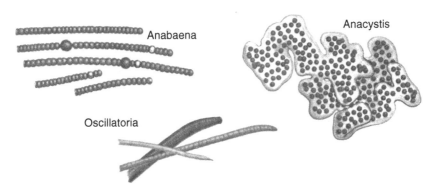

Figure 17-10 Blue Green Algae

(*Source:* Palmer, *Algae in Water Supplies,* U.S. Department of Health, Education, and Welfare, 1962.)

by fission (splitting). They prefer polluted, hard, and warm water. Some of them cause severe taste and odor problems. Common examples are *Anabaena, Oscillatoria,* and *Anacystis.*

- *Euglenophyta (euglenoides) are grass-green algae with flagella.* Euglenophytes are unicellular or colonies. They are found in dirty waters, such as ponds with decaying organic matter. Examples are *Euglena* and *Volvox* (see Figure 17-11).

- *Chlorophyta (green algae) are green unicellular and multicellular algae.* See Figure 17-12. Each genus has a distinct shape of chloroplast (chlorophyll body) in the cell. Examples are *chlorella,* and *Spirogyra* with cup-shaped and spiral-like chloroplasts, respectively. Some of them cause grassy odors.

- *Chrysophyta (golden brown and yellow-green algae) are golden brown or yellow-green algae with a rigid cell wall impregnated with silica, called shell.* Diatoms belong to this group (see Figure 17-13). The diatom shells, called frustules, form diatomaceous earth, which is used as filter media. They grow on the rocks in the cold and clean running streams. Some chrysophytes cause filter clogging. Examples are *Navicula, Synedra, Asterionella,* and *Staroneis.*

- *Pyrophyta (fire algae or dinoflagellates) are dark brown and unicellular algae with two flagella of unequal length.* Fresh water fire algae are present in highly polluted small ponds, rich in organic material, and during drought conditions.

- *Phaeophyta (brown algae) are large multicellular shallow water seaweeds like kelps.* Some of them are used as human food and food additives.

- *Rhodophyta (red algae) are red seaweeds.* They grow in deep oceans and some are used for human food. *Gelidium,* a rhodophyte, produces agar, which is used in semisolid culture media (semisolid below 45°C).

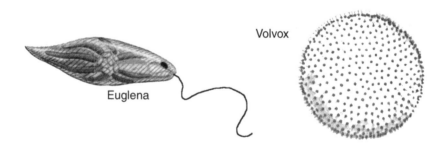

Figure 17-11 Euglena and Volvox

(*Source:* Palmer, *Algae in Water Supplies,* U.S. Department of Health, Education, and Welfare, 1962.)

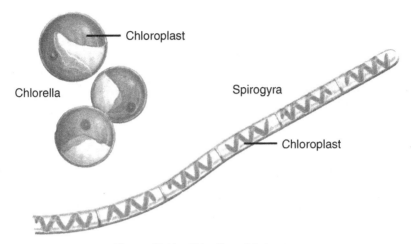

Figure 17-12 Chlorella and Spirogyra

(*Source:* Palmer, *Algae in Water Supplies,* U.S. Department of Health, Education, and Welfare, 1962.)

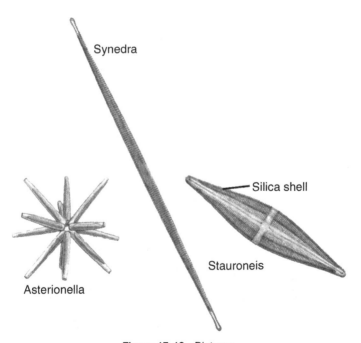

Figure 17-13 Diatoms

(*Source:* Palmer, *Algae in Water Supplies,* U.S. Department of Health, Education, and Welfare, 1962.)

Algal Control

As algae are photosynthetic, a small dose of powdered activated carbon blocks the sunlight and controls their growth in the reservoirs. The dose of 0.2 to 0.5 pounds/1,000 ft.² of water surface gives a good control. Other control measures are the use of copper sulfate (1 to 5 lb./acre) and 0.5 mg/L of free residual chlorine.

PROTOZOA

Protozoa are unicellular animals. An animal's ultimate source of food is plants. It differs from a plant by the absence of both chlorophyll and the cell wall. They are capable of moving around in search of food.

The active stage of protozoa, called trophozoite, has a flexible cell membrane (pellicle), which helps the body movements; dormant stage is a cyst.

Classification of Protozoa Based on the Means of Locomotion

- *Sarcodina (amoeboides)* move using *pseudopodia, false feet, which are formed at any place of the body as needed to move around and engulf food particles.* The term *amoeboide* is derived from the Greek word *amoibe meaning change because amoeboides change their shape often.* Pseudopodia are often withdrawn and reformed. Amoeboid body is like a balloon full of jelly, which can be squeezed in any form and in any direction. Common examples are *Amoeba proteus* and *Entamoeba histolytica* (see Figure 17-14). The latter causes the amebic dysentery, a waterborne disease.
- *Mastigophora (flagellates)* move in the liquid medium by flagella, which are long, whiplike extensions of the cytoplasm. *Giardia lamblia*, a waterborne pathogen, is a flagellate and causes giardiasis. *Giardia lamblia* has eight flagella in its active stage, trophozoite, which is pear-shaped with two nuclei at the wider end. It parasitizes the inner lining of the

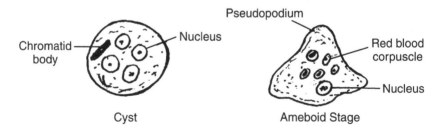

Figure 17-14   *Entamoeba histolytica*

large intestine of its host. The trophozoites form cysts, the dormant and infective stage. Cysts pass out with the fecal matter of the host and enter the surface waters. A *giardia* cyst is an oval structure, with two to four nuclei and a claw-hammer-shaped median body, enclosed in a tough and resistant cyst wall. The cyst is 5 to 15 micrometers wide and 8 to 18 micrometers long. Children under two years of age are more susceptible to this parasite.

G. lamblia's common host is the beaver. It is an opportunistic human parasite. Beavers pass this parasite with their fecal matter into rivers, streams, and lakes. The runoffs are the other means of its entering the surface waters. Improperly treated drinking water causes the infection in man. Other source of human infection is the contamination of food with fecal matter.

- *Ciliata (ciliophores) are more advanced protozoans with better means of locomotion, cilia, which are small, hairlike structures.* Wavelike synchronous movements of cilia cause fast-gliding movements of the organism. Common examples of this group are *Paramecium* and *Balantidium*.
- *Sporozoa are obligate parasites of higher animals, vertebrates and invertebrates.* They do not have any means of locomotion. *Cryptosporidium parvum,* a sporozoan, became the *super bug* of the 1990s due to the well-known Milwaukee, Wisconsin, cryptosporidiosis episode of 1993. Due to a combination of factors, the drinking water supply of the city of Milwaukee got contaminated with *Cryptosroridium* and made a large number of people ill. Currently, *Cryptosporidium* is the most important waterborne pathogen to control. It is commonly present in the large intestine of cattle. The active stage, trophozoite, multiplies in the inner lining of the large intestine of the host and produces the dormant stage, *oocyst* (fertilized egg or zygote). The oocyst, the infective stage, is a round structure about 3 to 7 micrometers in diameter, containing four banana-shaped sporozoites (see Figure 17-16). Oocysts are dis-

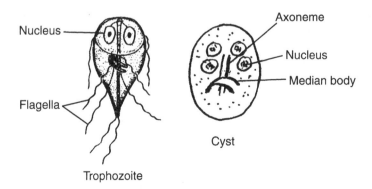

Figure 17-15 *Giardia lamblia*

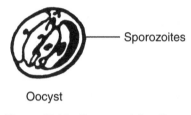

Oocyst

Figure 17-16 *Cryptosporidium* Oocyst

charged by the host with the fecal matter. After heavy rains, they get into streams through the runoffs from farm land or cattle feed lots.

When present in the potable water even in small numbers, *Cryptosporidium* causes cryptosporidiosis. This disease causes serious diarrhea and dehydration, especially in people with a weak immune system.

Protozoan Waterborne Pathogens

At present, only *Entameba histolytica, Giardia lamblia,* and *Cryptosporidium parvum* are protozoan pathogens of major concern.

Disease	Caused by
Amebic dysentery	*Entameba histolytica*
Giardiasis	*Giardia lamblia*
Cryptosporidiosis	*Cryptosporidium parvum*

The last two pathogens are hard to kill by common disinfection practices. The surface water treatment rule is particularly developed for the removal or destruction of these pathogens, by proper filtration and an effective disinfection plan.

All waterborne pathogens have two things in common: they are present in the human large intestine and cause gastrointestinal disorders.

Other microbes such as the *Cyclospora, Blastocystis,* and *microsporidia* group are emerging as possible future waterborne pathogens.

A basic knowledge of microbiology is essential for dealing with waterborne pathogens and for microbiological testing. We cannot fight an enemy effectively without knowing it and its habits adequately.

QUESTIONS

1. a. What are the major differences between living and nonliving things?

   b. Define *cell, nucleus, cytoplasm, cell wall,* and *cytoplasmic membrane.*

   c. An animal cell has chlorophyll and a cell wall. T or F

2. **a.** Viruses are the borderline case between living and nonliving things. T or F
   b. Viruses can multiply outside the host cell. T or F
   c. Enteric viruses are present in the large _____ of man.
   d. Name the well-known viral waterborne disease.

3. **a.** Bacteria are unicellular plants. T or F
   b. Bacteria reproduce asexually by binary _____.
   c. State three forms of bacteria and their technical names.
   d. Explain the terms *mesophilic, spore, binary fission,* and *generation time.*

4. **a.** Why is *E. coli* used as an indicator organism of human wastes?
   b. Differentiate between fecal and nonfecal coliform bacteria.
   c. Coliform bacteria are bacilli that ferment lactose to produce gas in 48 hrs at 35°C. T or F
   d. Write the characteristics of coliform group.
   e. Name four bacterial waterborne diseases.

5. **a.** How do fungi differ from algae?
   b. State various conditions that favor the growth of fungi.
   c. Give some common examples of fungi.

6. **a.** Blue-green algae mainly grow during summer in polluted ponds. T or F
   b. Diatoms are algae that normally indicate clean and running water. T or F
   c. Fresh-water fire algae indicate the stagnant water high in _____ matter.

7. **a.** Write characteristics of protozoa.
   b. Define *pseudopodium, flagellum,* and *cilia.*
   c. Name three important protozoan waterborne diseases.
   d. *Giardia* is a sporozoan and *Cryptosporidium* is a flagellate. T or F
   e. State the common hosts of *Giardia lamblia* and *Cryptosporidium parvum.*

8. Write a list of established waterborne diseases. What are their common characteristics?

APPENDIX A

INFORMATION SOURCES

Phone Numbers

- U.S. EPA hot line   (800) 426-4791
- American Water Works Association (AWWA)   (800) 926-7337

Web Sites
U.S. EPA

- Home page   www.epa.gov
- Safe Drinking Water Act (SDWA)   www.epa.gov/safewater/sdwa/
                                   index.html
- MCLs   www.epa.gov/safewater/
                                   mcl.html
- Regions   www.epa.gov/epahome/
                                   locate2.htm
- Water projects   www.epa.gov/epahome/
                                   waterpgram.htm

AWWA

- Home page   www.awwa.org
- Drinking water programs   www.awwa.org/stateinfo.htm

Federal Centers for Disease Control and Prevention

- Home page   www.cdc.gov
- *Cryptosporidium*   www.cdc.gov/ncidod/dpd/parasites/
                       cryptosporidiosis/default.htm
- *Giardia*   www.cdc.gov/ncidod/dpd/parasites/giardiasis/
                       default.htm

Besides these sources, contact your water utility and your state health department for the latest information on your drinking water.

APPENDIX B

CT VALUES

CT Values for Inactivation of *Giardia* Cysts by Chloramine pH 6-9

Inactivation	Temperature (°C)					
	<=1	5	10	15	20	25
0.5-log	635	365	310	250	185	125
1-log	1,270	735	615	500	370	250
1.5-log	1,900	1,100	930	750	550	375
2-log	2,535	1,470	1,230	1,000	735	500
2.5-log	3,170	1,830	1,540	1,250	915	625
3-log	3,800	2,200	1,850	1,500	1,100	750

CT Values for Inactivation of Viruses by Chloramine

Inactivation	Temperature (°C)					
	<=1	5	10	15	20	25
2-log	1,243	857	643	428	321	214
3-log	2,063	1,423	1,067	712	534	356
4-log	2,883	1,988	1,491	994	746	497

(*Source:* DoD 4715.5-G, "Overseas Environmental Baseline Guidance Document.")

CT Values for Inactivation of *Giardia* Cysts by Free Chlorine at 0.5°C or Lower*

Chlorine Concentration (mg/L)	pH <= 6 Log Inactivations						pH = 6.5 Log Inactivations						pH = 7.0 Log Inactivations						pH = 7.5 Log Inactivations					
	0.5	1.0	1.5	2.0	2.5	3.0	0.5	1.0	1.5	2.0	2.5	3.0	0.5	1.0	1.5	2.0	2.5	3.0	0.5	1.0	1.5	2.0	2.5	3.0
<=0.4	23	46	69	91	114	137	27	54	82	109	136	163	33	65	98	130	163	195	40	79	119	158	198	237
0.6	24	47	71	94	118	141	28	56	84	112	140	168	33	67	100	133	167	200	40	80	120	159	199	239
0.8	24	48	73	97	121	145	29	57	86	115	143	172	34	68	103	137	171	205	41	82	123	164	205	246
1	25	49	74	99	123	148	29	59	88	117	147	176	35	70	105	140	175	210	42	84	127	169	211	253
1.2	25	51	76	101	127	152	30	60	90	120	150	180	36	72	108	143	179	215	43	86	130	173	216	259
1.4	26	52	78	103	129	155	31	61	92	123	153	184	37	74	111	147	184	221	44	89	133	177	222	266
1.6	26	52	79	105	131	157	32	63	95	126	158	189	38	75	113	151	188	226	46	91	137	182	228	273
1.8	27	54	81	108	135	162	32	64	97	129	161	193	39	77	116	154	193	231	47	93	140	186	233	279
2	28	55	83	110	138	165	33	66	99	131	164	197	39	79	118	157	197	236	48	95	143	191	238	286
2.2	28	56	85	113	141	169	34	67	101	134	168	201	40	81	121	161	202	242	50	99	149	198	248	297
2.4	29	57	86	115	143	172	34	68	103	137	171	205	41	82	124	165	206	247	50	99	149	199	248	298
2.6	29	58	88	117	146	175	35	70	105	139	174	209	42	84	126	168	210	252	51	101	152	203	253	304
2.8	30	59	89	119	148	178	36	71	107	142	178	213	43	86	129	171	214	257	52	103	155	207	258	310
3	30	60	91	121	151	181	36	72	109	145	181	217	44	87	131	174	218	261	53	105	158	211	263	316

227

CT Values for Inactivation of *Giardia* Cysts by Free Chlorine at 0.5°C or Lower* (Continued)

Chlorine Concentration (mg/L)	pH <= 8 Log Inactivations						pH = 8.5 Log Inactivations						pH = 9.0 Log Inactivations					
	0.5	1.0	1.5	2.0	2.5	3.0	0.5	1.0	1.5	2.0	2.5	3.0	0.5	1.0	1.5	2.0	2.5	3.0
<=0.4	46	92	139	185	231	277	55	110	165	219	274	329	65	130	195	260	325	390
0.6	48	95	143	191	238	286	57	114	171	228	285	342	68	136	204	271	339	407
0.8	49	98	148	197	246	295	59	118	177	236	295	354	70	141	211	281	352	422
1	51	101	152	203	253	304	61	122	183	243	304	365	73	146	219	291	364	437
1.2	52	104	157	209	261	313	63	125	188	251	313	376	75	150	226	301	376	451
1.4	54	107	161	214	268	321	65	129	194	258	323	387	77	155	232	309	387	464
1.6	55	110	165	219	274	329	66	132	199	265	331	397	80	159	239	318	398	477
1.8	56	113	169	225	282	338	68	136	204	271	339	407	82	163	245	326	408	489
2	58	115	173	231	288	346	70	139	209	278	348	417	83	167	250	333	417	500
2.2	59	118	177	235	294	353	71	142	213	284	355	426	85	170	256	341	426	511
2.4	60	120	181	241	301	361	73	145	218	290	363	435	87	174	261	348	435	522
2.6	61	123	184	245	307	368	74	148	222	296	370	444	89	178	267	355	444	533
2.8	63	125	188	250	313	375	75	151	226	301	377	452	91	181	272	362	453	543
3	64	127	191	255	318	382	77	153	230	307	383	460	92	184	276	368	460	552

* $CT_{99.9}$ = CT for 3 log inactivation.

CT Values for Inactivation of *Giardia* Cysts by Free Chlorine at 5.0°C*

Chlorine Concentration (mg/L)	pH <= 6 Log Inactivations						pH = 6.5 Log Inactivations						pH = 7.0 Log Inactivations						pH = 7.5 Log Inactivations					
	0.5	1.0	1.5	2.0	2.5	3.0	0.5	1.0	1.5	2.0	2.5	3.0	0.5	1.0	1.5	2.0	2.5	3.0	0.5	1.0	1.5	2.0	2.5	3.0
<=0.4	16	32	49	65	81	97	20	39	59	78	98	117	23	46	70	93	116	139	28	55	83	111	138	166
0.6	17	33	50	67	83	100	20	40	60	80	100	120	24	48	72	95	119	143	29	57	86	114	143	171
0.8	17	34	52	69	86	103	20	41	61	81	102	122	24	49	73	97	122	146	29	58	88	117	146	175
1	18	35	53	70	88	105	21	42	63	83	104	125	25	50	75	99	124	149	30	60	90	119	149	179
1.2	18	36	54	71	89	107	21	42	64	85	106	127	25	51	76	101	127	152	31	61	92	122	153	183
1.4	18	36	55	73	91	109	22	43	65	87	108	130	26	52	78	103	129	155	31	62	94	125	156	187
1.6	19	37	56	74	93	111	22	44	66	88	110	132	26	53	79	105	132	158	32	64	96	128	160	192
1.8	19	38	57	76	95	114	23	45	68	90	113	135	27	54	81	108	135	162	33	65	98	131	163	196
2	19	39	58	77	97	116	23	46	69	92	115	138	28	55	83	110	138	165	33	67	100	133	167	200
2.2	20	39	59	79	98	118	23	47	70	93	117	140	28	56	85	113	141	169	34	68	102	136	170	204
2.4	20	40	60	80	100	120	24	48	72	95	119	143	29	57	86	115	143	172	35	70	105	139	174	209
2.6	20	41	61	81	102	122	24	49	73	97	122	146	29	58	88	117	146	175	36	71	107	142	178	213
2.8	21	41	62	83	103	124	25	49	74	99	123	148	30	59	89	119	148	178	36	72	109	145	181	217
3	21	42	63	84	105	126	25	50	76	101	126	151	30	61	91	121	152	182	37	74	111	147	184	221

CT Values for Inactivation of *Giardia* Cysts by Free Chlorine at 5.0°C* (Continued)

Chlorine Concentration (mg/L)	pH ≤ 8 Log Inactivations						pH = 8.5 Log Inactivations						pH = 9.0 Log Inactivations					
	0.5	1.0	1.5	2.0	2.5	3.0	0.5	1.0	1.5	2.0	2.5	3.0	0.5	1.0	1.5	2.0	2.5	3.0
≤=0.4	33	66	99	132	165	198	39	79	118	157	197	236	47	93	140	186	233	279
0.6	34	68	102	136	170	204	41	81	122	163	203	244	49	97	146	194	243	291
0.8	35	70	105	140	175	210	42	84	126	168	210	252	50	100	151	201	251	301
1	36	72	108	144	180	216	43	87	130	173	217	260	52	104	156	208	260	312
1.2	37	74	111	147	184	221	45	89	134	178	223	267	53	107	160	213	267	320
1.4	38	76	114	151	189	227	46	91	137	183	228	274	55	110	165	219	274	329
1.6	39	77	116	155	193	232	47	94	141	187	234	281	56	112	169	225	281	337
1.8	40	79	119	159	198	238	48	96	144	191	239	287	58	115	173	230	288	345
2	41	81	122	162	203	243	49	98	147	196	245	294	59	118	177	235	294	353
2.2	41	83	124	165	207	248	50	100	150	200	250	300	60	120	181	241	301	361
2.4	42	84	127	169	211	253	51	102	153	204	255	306	61	123	184	245	307	368
2.6	43	86	129	172	215	258	52	104	156	208	260	312	63	125	188	250	313	375
2.8	44	88	132	175	219	263	53	106	159	212	265	318	64	127	191	255	318	382
3	45	89	134	179	223	268	54	108	162	216	270	324	65	130	195	259	324	389

*$CT_{99.9}$ = CT for 3 log inactivation.

CT Values for Inactivation of *Giardia* Cysts by Free Chlorine at 10°C*

Chlorine Concentration (mg/L)	pH <= 6 Log Inactivations						pH = 6.5 Log Inactivations						pH = 7.0 Log Inactivations						pH = 7.5 Log Inactivations					
	0.5	1.0	1.5	2.0	2.5	3.0	0.5	1.0	1.5	2.0	2.5	3.0	0.5	1.0	1.5	2.0	2.5	3.0	0.5	1.0	1.5	2.0	2.5	3.0
<=0.4	12	24	37	49	61	73	15	29	44	59	73	88	17	35	52	69	87	104	21	42	63	83	104	125
0.6	13	25	38	50	63	75	15	30	45	60	75	90	18	36	54	71	89	107	21	43	64	85	107	128
0.8	13	26	39	52	65	78	15	31	46	61	77	92	18	37	55	73	92	110	22	44	66	87	109	131
1	13	26	40	53	66	79	16	31	47	63	78	94	19	37	56	75	93	112	22	45	67	89	112	134
1.2	13	27	40	53	67	80	16	32	48	63	79	95	19	38	57	76	95	114	23	46	69	91	114	137
1.4	14	27	41	55	68	82	16	33	49	65	82	98	19	39	58	77	97	116	23	47	70	93	117	140
1.6	14	28	42	55	69	83	17	33	50	66	83	99	20	40	60	79	99	119	24	48	72	96	120	144
1.8	14	29	43	57	72	86	17	34	51	67	84	101	20	41	61	81	102	122	25	49	74	98	123	147
2	15	29	44	58	73	87	17	35	52	69	87	104	21	41	62	83	103	124	25	50	75	100	125	150
2.2	15	30	45	59	74	89	18	35	53	70	88	105	21	42	64	85	106	127	26	51	77	102	128	153
2.4	15	30	45	60	75	90	18	36	54	71	89	107	22	43	65	86	108	129	26	52	79	105	131	157
2.6	15	31	46	61	77	92	18	37	55	73	92	110	22	44	66	87	109	131	27	53	80	107	133	160
2.8	16	31	47	62	78	93	19	37	56	74	93	111	22	45	67	89	112	134	27	54	82	109	136	163
3	16	32	48	63	79	95	19	38	57	75	94	113	23	46	69	91	114	137	28	55	83	111	138	166

CT Values for Inactivation of *Giardia* Cysts by Free Chlorine at 10°C* *(Continued)*

Chlorine Concentration (mg/L)	pH <= 8 Log Inactivations						pH = 8.5 Log Inactivations						pH = 9.0 Log Inactivations					
	0.5	1.0	1.5	2.0	2.5	3.0	0.5	1.0	1.5	2.0	2.5	3.0	0.5	1.0	1.5	2.0	2.5	3.0
<=0.4	25	50	75	99	124	149	30	59	89	118	148	177	35	70	105	139	174	209
0.6	26	51	77	102	128	153	31	61	92	122	153	183	36	73	109	145	182	218
0.8	26	53	79	105	132	158	32	63	95	126	158	189	38	75	113	151	188	226
1	27	54	81	108	135	162	33	65	98	130	163	195	39	78	117	156	195	234
1.2	28	55	83	111	138	166	33	67	100	133	167	200	40	80	120	160	200	240
1.4	28	57	85	113	142	170	34	69	103	137	172	206	41	82	124	165	206	247
1.6	29	58	87	116	145	174	35	70	106	141	176	211	42	84	127	169	211	253
1.8	30	60	90	119	149	179	36	72	108	143	179	215	43	86	130	173	216	259
2	30	61	91	121	152	182	37	74	111	147	184	221	44	88	133	177	221	265
2.2	31	62	93	124	155	186	38	75	113	150	188	225	45	90	136	181	226	271
2.4	32	63	95	127	158	190	38	77	115	153	192	230	46	92	138	184	230	276
2.6	32	65	97	129	162	194	39	78	117	156	195	234	47	94	141	187	234	281
2.8	33	66	99	131	164	197	40	80	120	159	199	239	48	96	144	191	239	287
3	34	67	101	134	168	201	41	81	122	162	203	243	49	97	146	195	243	292

* $CT_{99.9}$ = CT for 3 log inactivation.

232

CT Values for Inactivation of *Giardia* Cysts by Free Chlorine at 15°C*

Chlorine Concentration (mg/L)	pH <= 6 Log Inactivations						pH = 6.5 Log Inactivations						pH = 7.0 Log Inactivations						pH = 7.5 Log Inactivations					
	0.5	1.0	1.5	2.0	2.5	3.0	0.5	1.0	1.5	2.0	2.5	3.0	0.5	1.0	1.5	2.0	2.5	3.0	0.5	1.0	1.5	2.0	2.5	3.0
<=0.4	8	16	25	33	41	49	10	20	30	39	49	59	12	23	35	47	58	70	14	28	42	55	69	83
0.6	8	17	25	33	42	50	10	20	30	40	50	60	12	24	36	48	60	72	14	29	43	57	72	86
0.8	9	17	26	35	43	52	10	20	31	41	51	61	12	24	37	49	61	73	15	29	44	59	73	88
1	9	18	27	35	44	53	11	21	32	42	53	63	13	25	38	50	63	75	15	30	45	60	75	90
1.2	9	18	27	36	45	54	11	21	32	43	53	64	13	25	38	51	63	76	15	31	46	61	77	92
1.4	9	18	28	37	46	55	11	22	33	43	54	65	13	26	39	52	65	78	16	31	47	63	78	94
1.6	9	19	28	37	47	56	11	22	33	44	55	66	13	26	40	53	66	79	16	32	48	64	80	96
1.8	10	19	29	38	48	57	11	23	34	45	57	68	14	27	41	54	68	81	16	33	49	65	82	98
2	10	19	29	39	48	58	12	23	35	46	58	69	14	28	42	55	69	83	17	33	50	67	83	100
2.2	10	20	30	39	49	59	12	23	35	47	58	70	14	28	43	57	71	85	17	34	51	68	85	102
2.4	10	20	30	40	50	60	12	24	36	48	60	72	14	29	43	57	72	86	18	35	53	70	88	105
2.6	10	20	31	41	51	61	12	24	37	49	61	73	15	29	44	59	73	88	18	36	54	71	89	107
2.8	10	21	31	41	52	62	12	25	37	49	62	74	15	30	45	59	74	89	18	36	55	73	91	109
3	11	21	32	42	53	63	13	25	38	51	63	76	15	30	46	61	76	91	19	37	56	74	93	111

CT Values for Inactivation of *Giardia* Cysts by Free Chlorine at 15°C* (*Continued*)

Chlorine Concentration (mg/L)	pH <= 8 Log Inactivations						pH = 8.5 Log Inactivations						pH = 9.0 Log Inactivations					
	0.5	1.0	1.5	2.0	2.5	3.0	0.5	1.0	1.5	2.0	2.5	3.0	0.5	1.0	1.5	2.0	2.5	3.0
<=0.4	17	33	50	66	83	99	20	39	59	79	98	118	23	47	70	93	117	140
0.6	17	34	51	68	85	102	20	41	61	81	102	122	24	49	73	97	122	146
0.8	18	35	53	70	88	105	21	42	63	84	105	126	25	50	76	101	126	151
1	18	36	54	72	90	108	22	43	65	87	108	130	26	52	78	104	130	156
1.2	19	37	56	74	93	111	22	45	67	89	112	134	27	53	80	107	133	160
1.4	19	38	57	76	95	114	23	46	69	91	114	137	28	55	83	110	138	165
1.6	19	39	58	77	97	116	24	47	71	94	118	141	28	56	85	113	141	169
1.8	20	40	60	79	99	119	24	48	72	96	120	144	29	58	87	115	144	173
2	20	41	61	81	102	122	25	49	74	98	123	147	30	59	89	118	148	177
2.2	21	41	62	83	103	124	25	50	75	100	125	150	30	60	91	121	151	181
2.4	21	42	64	85	106	127	26	51	77	102	128	153	31	61	92	123	153	184
2.6	22	43	65	86	108	129	26	52	78	104	130	156	31	63	94	125	157	188
2.8	22	44	66	88	110	132	27	53	80	106	133	159	32	64	96	127	159	191
3	22	45	67	89	112	134	27	54	81	108	135	162	33	65	98	130	163	195

*$CT_{99.9}$ = CT for 3 log inactivation.

234

CT Values for Inactivation of Giardia Cysts by Free Chlorine at 20°C*

Chlorine Concentration (mg/L)	pH <= 6 Log Inactivations						pH = 6.5 Log Inactivations						pH = 7.0 Log Inactivations						pH = 7.5 Log Inactivations					
	0.5	1.0	1.5	2.0	2.5	3.0	0.5	1.0	1.5	2.0	2.5	3.0	0.5	1.0	1.5	2.0	2.5	3.0	0.5	1.0	1.5	2.0	2.5	3.0
<=0.4	6	12	18	24	30	36	7	15	22	29	37	44	9	17	26	35	43	52	10	21	31	41	52	62
0.6	6	13	19	25	32	38	8	15	23	30	38	45	9	18	27	36	45	54	11	21	32	43	53	64
0.8	7	13	20	26	33	39	8	15	23	31	38	46	9	18	28	37	46	55	11	22	33	44	55	66
1	7	13	20	26	33	39	8	16	24	31	39	47	9	19	28	37	47	56	11	22	34	45	56	67
1.2	7	13	20	27	33	40	8	16	24	32	40	48	10	19	29	38	48	57	12	23	35	46	58	69
1.4	7	14	21	27	34	41	8	16	25	33	41	49	10	19	29	39	48	58	12	23	35	47	58	70
1.6	7	14	21	28	35	42	8	17	25	33	42	50	10	20	30	39	49	59	12	24	36	48	60	72
1.8	7	14	22	29	36	43	9	17	26	34	43	51	10	20	31	41	51	61	12	25	37	49	62	74
2	7	15	22	29	37	44	9	17	26	35	43	52	10	21	31	41	52	62	13	25	38	50	63	75
2.2	7	15	22	29	37	44	9	18	27	35	44	53	11	21	32	42	53	63	13	26	39	51	64	77
2.4	8	15	23	30	38	45	9	18	27	36	45	54	11	22	33	43	54	65	13	26	39	52	65	78
2.6	8	15	23	31	38	46	9	18	28	37	46	55	11	22	33	44	55	66	13	27	40	53	67	80
2.8	8	16	24	31	39	47	9	19	28	37	47	56	11	22	34	45	56	67	14	27	41	54	68	81
3	8	16	24	31	39	47	10	19	29	38	48	57	11	23	34	45	57	68	14	28	42	55	69	83

CT Values for Inactivation of *Giardia* Cysts by Free Chlorine at 20°C* (Continued)

Chlorine Concentration (mg/L)	pH <= 8 Log Inactivations						pH = 8.5 Log Inactivations						pH = 9.0 Log Inactivations					
	0.5	1.0	1.5	2.0	2.5	3.0	0.5	1.0	1.5	2.0	2.5	3.0	0.5	1.0	1.5	2.0	2.5	3.0
<=0.4	12	25	37	49	62	74	15	30	45	59	74	89	18	35	53	70	88	105
0.6	13	26	39	51	64	77	15	31	46	61	77	92	18	36	55	73	91	109
0.8	13	26	40	53	66	79	16	32	48	63	79	95	19	38	57	75	94	113
1	14	27	41	54	68	81	16	33	49	65	82	98	20	39	59	78	98	117
1.2	14	28	42	55	69	83	17	33	50	67	83	100	20	40	60	80	100	120
1.4	14	28	43	57	71	85	17	34	52	69	86	103	21	41	62	82	103	123
1.6	15	29	44	58	73	87	18	35	53	70	88	105	21	42	63	84	105	126
1.8	15	30	45	59	74	89	18	36	54	72	90	108	22	43	65	86	108	129
2	15	30	46	61	76	91	18	37	55	73	92	110	22	44	66	88	110	132
2.2	16	31	47	62	78	93	19	38	57	75	94	113	23	45	68	90	113	135
2.4	16	32	48	63	79	95	19	38	58	77	96	115	23	46	69	92	115	138
2.6	16	32	49	65	81	97	20	39	59	78	98	117	24	47	71	94	118	141
2.8	17	33	50	66	83	99	20	40	60	79	99	119	24	48	72	95	119	143
3	17	34	51	67	84	101	20	41	61	81	102	122	24	49	73	97	122	146

* $CT_{99.9}$ = CT for 3 log inactivation.

CT Values for Inactivation of _Giardia_ Cysts by Free Chlorine at 25°C*

Chlorine Concentration (mg/L)	pH <= 6 Log Inactivations						pH = 6.5 Log Inactivations						pH = 7.0 Log Inactivations						pH = 7.5 Log Inactivations					
	0.5	1.0	1.5	2.0	2.5	3.0	0.5	1.0	1.5	2.0	2.5	3.0	0.5	1.0	1.5	2.0	2.5	3.0	0.5	1.0	1.5	2.0	2.5	3.0
<=0.4	4	8	12	16	20	24	5	10	15	19	24	29	6	12	18	23	29	35	7	14	21	28	35	42
0.6	4	8	13	17	21	25	5	10	15	20	25	30	6	12	18	24	30	36	7	14	22	29	36	43
0.8	4	9	13	17	22	26	5	10	16	21	26	31	6	12	19	25	31	37	7	15	22	29	37	44
1	4	9	13	17	22	26	5	10	16	21	26	31	6	12	19	25	31	37	8	15	23	30	38	45
1.2	5	9	14	18	23	27	5	11	16	21	27	32	6	13	19	25	32	38	8	15	23	31	38	46
1.4	5	9	14	18	23	27	6	11	17	22	28	33	7	13	20	26	33	39	8	16	24	31	39	47
1.6	5	9	14	19	23	28	6	11	17	22	28	33	7	13	20	27	33	40	8	16	24	32	40	48
1.8	5	10	15	19	24	29	6	11	17	23	28	34	7	14	21	27	34	41	8	16	25	33	41	49
2	5	10	15	19	24	29	6	12	18	23	29	35	7	14	21	27	34	41	8	17	25	33	42	50
2.2	5	10	15	20	25	30	6	12	18	23	29	35	7	14	21	28	35	42	9	17	26	34	43	51
2.4	5	10	15	20	25	30	6	12	18	24	30	36	7	14	22	29	36	43	9	17	26	35	43	52
2.6	5	10	16	21	26	31	6	12	19	25	31	37	7	15	22	29	37	44	9	18	27	35	44	53
2.8	5	10	16	21	26	31	6	12	19	25	31	37	8	15	23	30	38	45	9	18	27	36	45	54
3	5	11	16	21	27	32	6	13	19	25	32	38	8	15	23	31	38	46	9	18	28	37	46	55

CT Values for Inactivation of Giardia Cysts by Free Chlorine at 25°C* (Continued)

Chlorine Concentration (mg/L)	pH ≤ 8 Log Inactivations						pH = 8.5 Log Inactivations						pH = 9.0 Log Inactivations					
	0.5	1.0	1.5	2.0	2.5	3.0	0.5	1.0	1.5	2.0	2.5	3.0	0.5	1.0	1.5	2.0	2.5	3.0
<=0.4	8	17	25	33	42	50	10	20	30	39	49	59	12	23	35	47	58	70
0.6	9	17	26	34	43	51	10	20	31	41	51	61	12	24	37	49	61	73
0.8	9	18	27	35	44	53	11	21	32	42	53	63	13	25	38	50	63	75
1	9	18	27	36	45	54	11	22	33	43	54	65	13	26	39	52	65	78
1.2	9	18	28	37	46	55	11	22	34	45	56	67	13	27	40	53	67	80
1.4	10	19	29	38	48	57	12	23	35	46	58	69	14	27	41	55	68	82
1.6	10	19	29	39	48	58	12	23	35	47	58	70	14	28	42	56	70	84
1.8	10	20	30	40	50	60	12	24	36	48	60	72	14	29	43	57	72	86
2	10	20	31	41	51	61	12	25	37	49	62	74	15	29	44	59	73	88
2.2	10	21	31	41	52	62	13	25	38	50	63	75	15	30	45	60	75	90
2.4	11	21	32	42	53	63	13	26	39	51	64	77	15	31	46	61	77	92
2.6	11	22	33	43	54	65	13	26	39	52	65	78	16	31	47	63	78	94
2.8	11	22	33	44	55	66	13	27	40	53	67	80	16	32	48	64	80	96
3	11	22	34	45	56	67	14	27	41	54	68	81	16	32	49	65	81	97

* $CT_{99.9}$ = CT for 3 log inactivation.

CT Values for Inactivation of Viruses by Free Chlorine

Temperature (°C)	Log Inactivation 2.0		Log Inactivation 3.0		Log Inactivation 3.0	
	pH 6-9	pH 10	pH 6-9	pH 10	pH 6-9	pH 10
0.5	6	45	9	66	12	90
5	4	30	6	44	8	60
10	3	22	4	33	6	45
15	2	15	3	22	4	30
20	1	11	2	16	3	22
25	1	7	1	11	2	15

CT Values for Inactivation of *Giardia* Cysts by Chlorine Dioxide

	Temperature (°C)					
Inactivation	<=1	5	10	15	20	25
0.5-log	10	4.3	4	3.2	2.5	2
1-log	21	8.7	7.7	6.3	5	3.7
1.5-log	32	13	12	10	7.5	5.5
2-log	42	17	15	13	10	7.3
2.5-log	52	22	19	16	13	9
3-log	63	26	23	19	15	11

CT Values for Inactivation of Viruses by Chlorine Dioxide pH 6-9

	Temperature (°C)					
Removal	<=1	5	10	15	20	25
2-log	8.4	5.6	4.2	2.8	2.1	1.4
3-log	25.6	17.1	12.8	8.6	6.4	4.3
4-log	50.1	33.4	25.1	16.7	12.5	8.4

CT Values for Inactivation of Giardia Cysts by Ozone

	Temperature (°C)					
Inactivation	<=1	5	10	15	20	25
0.5-log	0.48	0.32	0.23	0.16	0.12	0.08
1-log	0.97	0.63	0.48	0.32	0.24	0.16
1.5-log	1.5	0.95	0.72	0.48	0.36	0.24
2-log	1.9	1.3	0.95	0.63	0.48	0.32
2.5-log	2.4	1.6	1.2	0.79	0.60	0.40
3-log	2.9	1.9	1.43	0.95	0.72	0.48

CT Values for Inactivation of Viruses by Free Ozone

Inactivation	Temperature (°C)					
	<=1	5	10	15	20	25
2-log	0.9	0.6	0.5	0.3	0.25	0.15
3-log	1.4	0.9	0.8	0.5	0.4	0.25
4-log	1.8	1.2	1.0	0.6	0.5	0.3

CT Values for Inactivation of Viruses by UV

Log Inactivation	
2.0	3.0
21	36

APPENDIX C

CHEMICALS USED IN WATER TREATMENT

Chemical Name	Common Name	Chemical Formula	Used for
Aluminum oxide	Liquid alum	Al_2O_3	Coagulation and defluoridation
Aluminum sulfate	Filter alum	$Al_2(SO_4) \cdot 14H_2O$	Coagulation
Ammonia	Ammonia gas	NH_3 (ammonia gas)	
	Ammonia aqua liquid	NH_4OH (ammonia solution)	Chloramination
Calcium bicarbonate		$Ca(HCO_3)_2$	Alkalinity
Calcium carbonate	Limestone	$CaCO_3$	
Calcium hydroxide	Hydrated lime or slaked lime	$Ca(OH)_2$	Softening
Calcium hypochlorite	HTH	$Ca(ClO)_2$	Chlorination
Calcium oxide	Unslaked lime or quick lime	CaO	Softening
Carbon	Activated carbon	C	Taste, odor, and pesticide removal
Chlorine		Cl_2	Chlorination
Chlorine dioxide		ClO_2	Disinfection
Copper sulfate	Blue vitriol	$CuSO_4 \cdot 5HO$	Algae control
Ferric chloride		$FeCl_3 \cdot 6H_2O$	Coagulation
Ferric sulfate		$Fe_2(SO_4)_3$	Coagulation
Ferrous chloride		$FeCl_2$	Chlorite control
Fluosilicic acid (hydrofluosilicic acid)	Fluoride	$H_2Si_2F_6$	Fluoridation

(*Source: Basic Chemistry for Water and Wastewater Operators,* by permission. Copyright © 2002, American Water Works Association.)

Chemical Name	Common Name	Chemical Formula	Used for
Hydrochloric acid	Muriatic acid	HCl	Acidity
Ozone		O_3	Disinfection
Potassium permanganate		$KMnO_4$	Taste and odor control
Sodium aluminate		$NaAlO_2$	Coagulation softening
Sodium bicarbonate	Baking soda	$NaHCO_3$	Alkalinity
Sodium carbonate	Soda ash	Na_2CO_3	Softening
Sodium chlorite		$NaClO_2$	Chlorine dioxide formation
Sodium fluoride		NaF	Fluoridation
Sodium fluosilicate		Na_2SiF_6	Fluoridation
Sodium hexametaphosphate	Calgon	$Na_6(PO_3)_6$ or $6(NaPO_3)$	Sequestering
Sodium hydroxide	Lye	NaOH	Alkalinity
Sodium hypochlorite	Bleach	NaClO	Chlorination
Sodium phosphate		$Na_3PO_4 \cdot 12H_2O$	
Sodium thiosulfate		$Na_2S_2O_3$	
Sulfuric acid	Oil of vitriol	H_2SO_4	
Zinc orthophosphate		$Zn_3(PO_4)_2$	Corrosion control

APPENDIX D

CHEMICAL REACTIONS
IN WATER TREATMENT

Coagulation and Flocculation

1. $Al_2O_3 + 3H_2O \rightarrow 2Al(OH)_3\downarrow$
2. $Al_2(SO_4)_3 + 3Ca(OH)_2 \rightarrow 2Al(OH)_3\downarrow + 3CaSO_4$
3. $Al_2(SO_4)_3 + 3Ca(HCO_3)_2 \rightarrow 2Al(OH)_3\downarrow + 3CaSO_4 + 6CO_2$
4. $Al_2(SO_4)_3 + 3Na_2CO_3 + 3H_2O \rightarrow 2Al(OH)_3\downarrow + 3Na_2SO_4 + 3CO_2$
5. $Fe_2(SO_4)_3 + 3Ca(OH)_2 \rightarrow 2Fe(OH)_3\downarrow + 3CaSO_4$

Softening and Stabilization

6. $Ca(HCO_3)_2 + Ca(OH)_2 \rightarrow 2CaCO_3\downarrow + 2H_2O$
7. $CaSO_4 + Na_2CO_3 \rightarrow CaCO_3\downarrow + Na_2SO_4$
8. $Mg(HCO_3)_2 + Ca(OH)_2 \rightarrow CaCO_3\downarrow + MgCO_3 + 2H_2O$
9. $MgCO_3 + Ca(OH)_2 \rightarrow CaCO_3\downarrow + Mg(OH)_2\downarrow$
10. $MgSO_4 + Ca(OH)_2 \rightarrow CaSO_4 + Mg(OH)_2\downarrow$

Iron and Manganese Removal

11. $Fe(HCO_3)_2 \rightarrow FeO + 2CO_2 + H_2O$
12. $4Fe(HCO_3)_2 + 8Ca(OH)_2 + O_2 \rightarrow 2Fe_2O_3\downarrow + 8CaCO_3\downarrow + 12H_2O$
13. $4FeO + O_2 \rightarrow 2Fe_2O_3\downarrow$ (Rust)
14. $2FeO + HOCl \rightarrow Fe_2O_3\downarrow + HCl$

(*Source: Basic Chemistry for Water and Wastewater Operators,* by permission. Copyright ©
2002, American Water Works Association.)

15. $2Mn(HCO_3)_2 + 4Ca(OH)_2 + O_2 \rightarrow 2MnO_2\downarrow + 4CaCO_3\downarrow + 6H_2O$
16. $Mn(HCO_3)_2 \rightarrow MnO + 2CO_2 + H_2O$
17. $6MnO + O_2 \rightarrow 2Mn_3O_4$ (an unstable compound)
18. $4Mn_3O_4 + O_2 \rightarrow 6Mn_2O_3\downarrow$
19. $3MnO + HOCl \rightarrow Mn_3O_4 + HCl$
20. $2Mn_3O_4 + HOCl \rightarrow 3Mn_2O_3\downarrow + HCl$

Disinfection

21. $Cl_2 + H_2O \rightarrow HCl + HOCl$
22. $Ca(OCl)_2 + 2H_2O \rightarrow Ca(OH)_2 + 2HOCl$
23. $CaClClO + 2H_2O \rightarrow CaCl_2 + Ca(OH)_2 + 2HOCl$
24. $NaClO + H_2O \rightarrow NaOH + HOCl$
25. $NH_3 + HOCl \rightarrow NH_2Cl + H_2O$
26. $NH_2Cl + HOCl \rightarrow NHCl_2 + H_2O$
27. $2NH_2Cl + HOCl \rightarrow N_2\uparrow + 3HCl + H_2O$
28. $NH_3 + 2HOCl \rightarrow NHCl_2 + 2H_2O$
29. $NHCl_2 + HOCl \rightarrow NCl_3 + H_2O$
30. $4KClO_3 + 4HCl \rightarrow 4KCl + 4ClO_2\uparrow + 2H_2O + O_2$
31. $2NaClO_2 + Cl_2 \rightleftarrows 2NaCl + 2ClO_2\uparrow$
32. $5NaClO_2 + 4HCl \rightleftarrows 5NaCl + 4ClO_2\uparrow + 2H_2O$
33. $3O_2 \rightleftarrows 2O_3$
34. $O_3 \rightarrow O_2 + O$

Neutralization

35. $CO_2 + Ca(OH)_2 \rightarrow CaCO_3\downarrow + H_2O$, above pH 9.4
36. $CO_2 + H_2O + CaCO_3 \rightarrow Ca(HCO_3)_2$, below pH 9.4
37. $CaCO_3 + H_2SO_4 \rightarrow CaSO_4 + H_2O + CO_2$
38. $Ca(HCO_3)_2 + H_2SO_4 \rightarrow CaSO_4 + 2H_2O + 2CO_2$
39. $Ca(OH)_2 + Na_2CO_3 \rightarrow CaCO_3\downarrow + 2NaOH$

Compounds Causing Acidity and Alkalinity in Water

40. $CO_2 + H_2O \rightarrow H_2CO_3$
41. $Al_2(SO_4)_3 + 3H_2O \rightarrow 2Al(OH)_3\downarrow + 3H_2SO_4$
42. $CaO + H_2O \rightarrow Ca(OH)_2$

GLOSSARY

Accelator. Type of a solid contact basin for sedimentation.

Acid. A compound that forms hydronium ions in water solution. It is a proton donor.

Actinomycetes. Branched bacteria that decompose organic matter and cause peculiar earthy-musty odors in the water.

Activated carbon. Charcoal with adsorbing pores and crevices created at a high temperature.

Activation. Creating pores and crevices in the charcoal at high temperature in the presence of steam.

Adsorbant. Substance that adsorbes other substances.

Adsorbate. Substance that is adsorbed.

Adsorption. The acquisition/accumulation of a gas, liquid, or solid on the surface of a solid particle.

Agar. Extract of seaweeds, a solidifying agent, which is used in the media to facilitate the growth of individual microbes.

Aggressive water. Very corrosive water.

Algae. Lower plants, unicellular and multicellular, with chlorophyll.

Amphiprotic. Capable of acting either as an acid or as a base.

Angstrom. A unit of linear measure; 10^{-8} cm.

Anhydrous. Without water of crystallization.

Anion. A negative ion.

Anode. Electrode that attracts anions. Positively charged pole. Electrode where oxidation takes place.

Aquifer. Sandy layers of soil saturated with water with an impervious stratum underneath.

Artesian wells. Wells that use artesian aquifer (aquifer in between two impervious strata).

Atmospheric water. Water vapor in the atmosphere that forms clouds.

Atom. The smallest particle of an element, capable of entering into combinations with other elements.

Atomic number. The number of protons in the nucleus of an atom.

Atomic weight. Relative average atomic mass of an element compared to one twelfth the mass of carbon-12 isotope.

Avogadro number. Number of carbon-12 atoms in exactly 12 grams of this isotope; 6.022169×10^{23}.

Base. A substance that accepts protons from another substance.

Binary compounds. Compounds that are formed of two elements.

Bio-film. Thin biological slimy covering in the pipes of the distribution system.

Blue vitriol. Hydrate copper (II) sulfate, $CuSO_4 \cdot 5H_2O$.

Booster pump. Pump for boosting pressure in the distribution.

Bored wells. Wells that are excavated with hand or power auger.

Buffer. A substance that resists any change in pH.

Butterfly valve. Disclike plate rotating around an axis to open or close the valve to any degree.

Calorie. The quantity of heat required to raise the temperature of 1 gram of water by 1°C.

Carbonation. Burning wood at about 700°C in the absence of air.

Catalyst. Substance that alters the rate of a chemical reaction but remains unchanged itself at the end of the reaction.

Cathode. Negatively charged electrode. It attracts cations. Electrode where reduction occurs.

Cation. Positively charged ion.

Cavitation. Formation of air pockets due to fusion of small bubbles in the pipes.

Centrifugal pump. Pump with a disc that rotates in a casing, sucks water, and slings it out by the centrifugal force.

Chemical bond. Bondage between atoms produced by transfer or sharing of electrons.

Chemical change. A change in which new chemicals with new properties are formed.

Chlorophyta (green algae). Green unicellular and multicellular algae. Their cell wall has cellulose.

Chrysophyta (diatoms, golden brown and yellow-green algae). Golden brown or yellow-green algae with silica shells.

Ciliata (ciliophores). Protozoans with cilia, hairlike means of locomotion.

Coagulant Aids. Substances that help the coagulants by creating better conditions.

Coagulant. Chemical such as alum that provides cations (Al^{+3}) to precipitate out turbidity.

Coagulation. Precipitation of the colloidal turbidity particles by coagulants.

Colloidal suspension. Suspended particles (1–100 nanometers in diameter) in a dispersing medium.

Compound. Chemical, which is composed of two or more elements combined in a definite proportion.

Concentrated. Containing a large amount of solute.

Cone of depression. An inverted conelike dewatered area around the well while pumping.

Conventional treatment plant. A typical municipal water treatment plant with sedimentation, filtration, and disinfection.

Covalence. Covalent bonding.

Covalent bond. Bond formed by a pair of shared electrons.

CT. Stands for concentration of a disinfectant as mg/L multiplied by the contact time in minutes.

Cyanophyta (blue-green algae). Unicellular or multicellular blue-green algae. Chlorophyll is diffused in the cytoplasm.

Defluoridation. Removal of excessive amount of fluoride from the water.

Dehydration. Removal of water from a substance.

Density. The mass per unit volume of a substance.

Dental fluorosis. Tooth disease (mottled teeth) caused by excessive amount of fluoride in the water.

Desorption. Release of adsorbed molecules.

Dialysis. Separation of dissolved substances from colloids/suspended matter in a liquid through a permeable membrane.

Diatomaceous earth filters. Pressure filters that use diatomaceous earth instead of sand.

Diatomic. Particle consisting of two atoms.

Diffusion. Process of spreading out of particles to fill a space uniformly.

Dilute. Containing a small amount of dissolved solute.

Dissociation. The separation of ions of an ionic solute during the solution formation.

Draw down. Difference between the static level and the pumping level.

Drilled wells. Deep drilled wells that are used by the water utilities.

Driven wells. Wells constructed by driving a pipe, with a pointed screen attached to its end, into the aquifer.

Dual media filters. Filters with two media, sand and anthracite.

Dug wells. Wells of 5 to 40 feet in diameter, dug with hand tools.

Effluent/filtrate. Filtered water (filter discharge).

Electrochemical. Pertaining to spontaneous oxidation reduction reactions used as a source of electric energy.

Electrodialysis. Separation of dissolved electrolytes from the water. Anions are collected at anode and cations at cathode after passing through a resin membrane.

Electrolysis. Decomposition of a substance by electricity.

Electrolyte. Substance whose water solution conducts electricity.

Electron. Negatively charged particle revolving around the nucleus of an atom.

Electrovalence. Ionic bonding.

Element. Substance that cannot be decomposed by ordinary chemical means.

End point or equivalence point. Point in a titration at which quantities of the standard and standardized chemicals are chemically equivalent.

Energy level. Region around the nucleus of an atom in which electrons revolve.

Enteric bacteria. Bacteria present in the large intestine.

Enzyme. A catalyst produced by the living cells.

Epilimnion. Top and the lightest layer of water in a stratified lake.

Equilibrium. A dynamic state in which two opposing processes proceed at the same rate at the same time.

Euglenophyta (euglenoides). Grass-green algae with flagella (whiplike outgrowths), unicellular or colonies.

Eutrophic lake. Old and shallow lake due to high amount of nutrients; has large algal blooms.

Eutrophication. Natural aging process of a lake.

Evaporation. Escape of molecules from the surface of liquids and solids.

Fecal coliform bacteria. *E. coli* and some other closely related enteric bacteria.

Filter cycle. Period between two backwashings.

Filter run. Time period from the start of filtration up to the filter backwashing.

Filtration. Removal of suspended particles mechanically by passing the water through a porous medium.

Flocculation. Floc formation by gentle mixing of coagulated water.

Flow. Quantity of water flowing per unit time.

Fluoridation. Use of fluorides in the drinking water.

Force. Weight on the bottom of a column of water.

Formula. Shorthand representation of the composition of a chemical by using chemical symbols and numerical subscripts.

Fouling. Plugging up of the membrane due to accumulation of particles.

Freezing point. Temperature at which a liquid becomes a solid.

Friction head. Loss of pressure due to friction.

Fungi. Lower plants (body is not divided into root, stem, and leaves), without chlorophyll.

Gas. State of matter in which a substance does not possess a definite shape or volume.

Gram. Metric unit of mass equal to the mass of 1 milliliter of water at 4°C.

Gram-atomic weight. Mass in grams of one mole of naturally occurring atoms of an element.

Granular activated carbon (GAC). Activated carbon in granular form.

Groundwater. Subterranean/underground water.

Growth media. Media—the food—for the microbes to culture them in the laboratory.

Head loss. Difference between the pressure on the influent and effluent sides of the filter water.

High rate sand filters. Modification of the rapid sand filters with 5 to 10 gpm/ft.^{2} loading.

High service pump. Pump to transport water under high pressure from the treatment plant to the distribution.

Hollow fine fiber membrane. A hollow fine fiber tubular membrane.

Homogeneous. Having uniform properties throughout.

Hydrate. A crystallized substance with water of crystallization.

Hydration. Association of water molecules to particles of the solute.

Hydraulics. Science of fluids.

Hydrologic (water) cycle. Cycle in which atmospheric, surface, and ground water are circulating.

Hydrolysis. A chemical reaction in which water is involved for decomposition of a substance.

Hydronium ion. A hydrated proton (hydrogen ion), H_3O^+ ion.

Hypolimnion. Bottom (the heaviest) layer of water in a stratified lake.

Impeller. Rotating disc of a centrifugal pump with vanes of various shapes.

Indicator organism. Organism that indicates the presence of other specific organisms.

Indicator. Substance that changes in color from a standard reagent to the standardized.

Infiltration. Soaking of rain water through the soil.

Ion. Atom or group of atoms with an electric charge.

Ionic bonding. Chemical bonding in which electrons are transferred from one atom to an other.

Ionization. Formation of ions from polar solute molecules by the action of polar molecules of the solvent.

Joule. Unit of energy that is equal to 0.239 calories.

Lentic (lenis = calm). Calm waters, such as lakes and reservoirs.

Lime. Calcium oxide, CaO, also known as quick lime; and $Ca(OH)_2$, slaked lime.

Liquid. State of matter that has a definite volume but no definite shape.

Liter. Volume occupied by 1 kilogram of water at 4°C.

Lotic (lotus means washed). Running waters such as rivers and streams.

Lye. Commercial grade of sodium hydroxide or potassium hydroxide.

Mass number. Total of the number of protons and neutrons in an atom.

Mastigophora (flagellates). Protozoans with flagella, whiplike structures used for locomotion.

MCL. Maximum contaminant level of a health-affecting substance.

MCLG. Maximum contaminant level goal of a contaminant is a level with no known adverse effects.

Melting point. A temperature at which a solid changes into a liquid.

Membrane filtration. Passing of water through a membrane to remove a specific size of particles, both suspended and dissolved.

Membrane fouling. Clogging of a membrane with filtered-out particles.

Mesophilic bacteria. Bacteria that grow at intermediate temperature.

Mesotrophic lakes. Middle-aged lake due to nutrients and sediments being added.

Meter. Metric unit of length. It is equal to 39.37 inches (1/10,000,000 of the North Polar quadrant of the Paris Meridian).

Microfiltration membranes. Membranes with pore size ranges 0.1 to 1 micrometer.

Mixed flow. Flow of a pump that is at a 45-degree angle to the suction.

Molality. An expression of the number of moles of a solute per kilogram of solvent.

Molarity. Expression of the moles of solute per liter of solution. It is also called *formality*.

Mole. Amount of a substance in grams that contains an Avogadro number of its particles. Practically, it is gram atomic weight for an element and gram formula weight of a compound.

Molecular formula. A chemical formula representing the composition of a molecule of a substance.

Molecular weight. Formula weight of a molecular substance.

Molecule. The smallest stable and neutral unit of a substance.

Monoprotic. Acid capable of donating one proton per molecule.

Nanofiltration. Membrane filters that remove particle of nanometer size (0.001 micrometer or 1 nanometer).

Neutralization. Reaction between a base and an acid to produce a salt and water.

Neutron. Neutral particle in an atomic nucleus with almost the same mass as that of a proton.

Nonelectrolyte. Substance that does not conduct electricity in its water solution.

Normal solution. Solution containing 1 gram equivalent weight of a solute per liter of solution.

Normality. Expression of the number of gram equivalent weights of a solute per liter of solution.

Nucleus. Positively charged and centrally located part of an atom.

Oligotrophic lake. Young, deep, and clear lake, with very little amount of nutrients.

Organic. Regarding carbon compounds and their derivatives.

Osmosis. Passage of water from the lower concentration toward the higher of dissolved substances, through a semipermeable membrane.

Oxidation number. Number of electrons of an element participating in compound formation. It is equal to electrons in neutral atom minus electrons in combined atoms.

Oxidation. Chemical reaction in which electrons are lost.

Oxidation-reduction reaction. Chemical reaction in which electrons are transferred.

Oxidizing agent. Substance that gains electrons in a chemical reaction.

Percolation. Deeper movement of water through the pores between sand particles.

Period. Horizontal row of elements in the Periodic Table.

Permanent hardness (noncarbonate hardness). Hardness in water due to the presence of sulfates, nitrates, and chlorides of calcium and magnesium.

Permeate. Filtrate or effluent of the membrane filter.

Peroxone. Combination of hydrogen peroxide and ozone for disinfection.

pH. Minus logarithm of the hydronium ion concentration, expressed as moles per liter.

Phaeophyta (brown algae). Large multicellular shallow-water marine seaweeds such as kelps.

Physical change. Change in which chemical composition of a substance remains unchanged.

Polarization. Accumulation of hydrogen gas and hydroxide ions on the cathode in a corrosion process.

Polarographic. Automatic measuring and recording system.

Precipitation. 1. Condensation of water vapors into rain, snow, and sleet. 2. Separation of a solid from a solution.

Pressure filters. Rapid sand filters enclosed in a cylindrical steel tank and operated under pressure rather than gravity.

Pressure. Force per unit of bottom area.

Pretreatment. Preparation of raw water for an adequate treatment by removing any large floating objects such a sticks, leaves and logs, fish, rags; and settlable solids like sand and other debris and includes partial disinfection.

Proton. Positively charged particle in the nucleus.

Protozoa. Unicellular lower animals.

Psychrophilic bacteria. Bacteria that prefer low temperature.

Pump. Device for transportation of a liquid.

Pyrophyta (fire algae or dinoflagellates). Dark brown, unicellular algae with cellulose in the cell wall and two flagella of unequal length.

Radial flow. Flow of a pump at right angle to the suction.

Radical. A group of covalently bonded atoms carrying a charge.

Radius of influence. Radius of the circular area of aquifer that is dewatered around the well.

Rapid mixing. Fast and thorough (flash) mixing of various chemicals into the water.

Rapid sand filters. Filters with surface loading 2 to 4 gpm/ft.^{2}

Reciprocating pumps. Pump with a mechanism to fill a chamber with liquid and to discharge it.

Redox. Regarding reduction-oxidation reactions.

Reducing agent. Substance that loses electrons in a chemical reaction.

Reduction. Any reaction in which electrons are gained.

Regeneration. Reactivation of an exhausted granular activated carbon.

Reverse osmosis membrane. Semipermeable membrane that allows passage of water and only some specific solutes (dissolved substances) to pass through.

Reversible reaction. A chemical reaction in which the products reform the reactants.

Rhodophyta (red algae). Red seaweeds. They grow in deep oceans.

Rotary pumps. A pump with two cams or gears; one is connected to the shaft and is called the driving gear that drives the other, the idler one.

Salt. An ionic compound formed by the reaction of an acid and a base.

Sarcodina. Ameboid protozoans. Means of locomotion are pseudopodia.

Saturated solution. Solution in which rate of dissolving is equal to the rate of crystallizing.

Sedimentation basins (clarifiers). Large basins for settling of solids by the force of gravity.

Sequestering. Keeping calcium and iron in solution by using a small dose of a polyphosphate such as sodium hexametaphosphate.

Shaft. Steel, stainless steel, or bronze rod of a pump to which impeller is attached.

Shell. Part of an atom around the nucleus of an atom in which electrons revolve.

Short circuits. Flowing of water through a basin without complete mixing.

Silt density index (SDI). Index used to determine the fouling degree of the water for membrane filtration.

Skeletal fluorosis. Bone disease (brittle bones) caused by the excessive amount of fluorides in water.

Slaked lime. Calcium hydroxide, $Ca(OH)_2$.

Slow sand filters. Sand filter with loading of less than $1 gpm/ft.^2$

Sludge blanket. Cloud like suspension of solids in a solid contact basin.

Soft water. Water with low calcium and magnesium contents.

Solute. Dissolved substance of a solution.

Solution equilibrium. Physical state in solution at which rate of crystallizing is equal to the rate of dissolving.

Solution. Homogeneous mixture of two or more substances.

Solvent. Dissolving medium of a solution.

Soozone. Combination of ultrasonic waves and ozone for disinfection.

Specific capacity. Capacity (water production) of a well per foot drawdown.

Specific gravity. Ratio of the density of a substance to the density of a standard. Standard for solids and liquids is water with the density of $1 g/cm^3$; for gases, the standard is air with the density of $1.29 g/L$ at STP.

Spiral wound membrane. Membrane formed of a flat sheet rolled around a center tube.

Spore. Resistant stage of a microorganism for unfavorable conditions.

Sporozoa. Protozoans that are obligate (strictly) parasites of higher animals, invertebrates and vertebrates.

Standard solution. Solution with the precise concentration of the solute.

Static level. Height to which the water rises when not pumping.

Stoichiometry. Qualitative relationship between reactants and products in a chemical reaction.

STP. Standard temperature ($0°C$) and pressure (760 mm Hg).

Stratification. Layering of water in lakes and ponds during summer and winter.

Suction head. Height between the free surface of the water and the center line of a pump.

Suction lift. Distance between the free surface of the source water and the center line of a pump.

Super saturated. Solution that contains more dissolved solute than a saturated solution would contain under similar conditions.

Surface water. Water on the surface of earth such as oceans, lakes, and rivers.

Temperature. Measure of the transferable heat of a system.

Temporary hardness (carbonate hardness). Hardness caused by bicarbonates of calcium and magnesium.

Thermocline. Middle layer of water in a stratified lake.

Thermophilic bacteria. Bacteria that prefer high temperatures—above 45°C.

Threshold odor number. Dilution factor with the least detectable odor.

Titration. Technique by which the concentration of a reagent is determined, by reacting it with a reagent of known concentration.

Total dynamic head. Total static head plus friction head.

Total static head. Vertical distance, from the free surface of the source water and the free surface of discharged water.

Transpiration. Evaporation of water from stomata of the plant leaves.

Triple media (mixed media) filter. Modification of dual media filters by adding a third layer of a heaviest medium like garnet, below the sand.

Turbidity. Murkiness in water caused by colloidal and other suspended particles.

Ultrafiltration. Membrane filtration with membranes of 0.003 to 0.1 micrometers pore size.

Valence electrons. Electrons in the outermost shell.

Valve. Device to control the direction and amount of flow of a liquid.

Vapor. Gaseous state of substances that normally exist as solids or liquids.

Velocity head. Pressure due to velocity.

Volatile. Easily vaporized.

Walker solid contact basin. Type of a solid contact sedimentation basin.

Water hammer. Surge or thrust of water pressure caused by a sudden stopping of the flow.

Water table wells. Wells that use water table aquifer.

Well pump. Pump to lift water from the well and discharge it to the treatment plant.

Zeta potential. Magnitude of the charge at the boundary between the colloidal particle and medium (water).

REFERENCES

Adham, S. S., J. C. Jacangelo, and J. M. Laine. 1995. "Low-Pressure Membrane; Assessing Integrity." *Journal of AWWA* 87 (3): 62–75.

American Water Works Association. 1978. *Introduction to Water Sources and Transmission.* Denver, Colo.: AWWA.

American Water Works Association. 1964. *Simplified Procedures for Water Examination.* Denver, Colo.: AWWA M12.

American Water Works Association. 1982. *Water Quality Analysis.* Denver, Colo.: AWWA.

American Water Works Association. 1995. *Problem Organisms in Water.* Denver, Colo.: AWWA M7.

American Water Works Association. 1999. *Water Quality and Treatment, A Handbook of Community Water Supplies.* McGraw Hill Handbooks.

American Water Works Association. 1999. *Manual of Water Supply Practices Reverse Osmosis and Nanofitration.* Denver, Colo.: AWWA AM46.

American Water Works Association Research Foundation. 1996. *Internal Corrosion of water Distribution Systems.* Denver, Colo.: AWWA and AWWaRF.

American Water Works Association. 1982. *Treatment Techniques for Controlling Trihalomethanes in Drinking Water.* Denver Colo.: AWWA.

Baylis, J. R. 1930. "How to Avoid Loss by Pipe Corrosion by Water." *Water Works Engineering* 83.

Cotton, C. A., D. M. Owen, G. C. Cline, and T. P. Brodeur. 2001. "UV Disinfection Costs for Inactivating Cryptosporidium." *Journal of AWWA* 93 (6): 82–94.

Finch, G. R., L. R. Liyanga, and M. Belosevic. 1995. *Effect of Chlorine Dioxide on Cryptosporidium and Giardia. Chlorine Dioxide: Drinking Water, and Wastewater Issues.* Third International Symposium.

Griffin, A. E., and R. J. Baker. 1959. "The Break-point Process for Free Residual Chlorination." *Journal of the New England Water Works Association.*

Hurst, G. H., and W. R. Knocke. 1997. "Evaluating Ferrous Iron for Chlorite Ion Removal." *Journal of AWWA* 89:8:98.

Langelier, W. F. 1936. "The Analytical Control of Anticorrosion, Water Treatment." *Journal AWWA:* 28.

Mofidi, A. A., H. Baribeau, P. A. Rochelle, R. D. Leon, and B. M. Coffey. 2001. "Disinfection of *Cryptosporidium parvum* with polychromatic UV Light." *Journal AWWA* 93 (6): 95–109.

New York State Department of Health. *Manual of Instruction for Water Treatment Plant Operators.* Albany, New York.

Pizzi, N. G. 2002. *Water Treatment Operator Handbook.* Denver Colo.: AWWA.

Sarai, D. S. 2002. *Basic Chemistry for Water and Wastewater Operators.* Denver Colo.: AWWA.

Sarai, D. S. 1976. "Total and Fecal Coliform Bacteria in Some Aquatic and Other Insects." *Environ. Entomol.,* 5: 365–367.

Tillman, G. M. 1996. *Water Treatment Troubleshooting and Problem Solving.* Boca Raton, London, New York, Washington, D.C.: Lewis Publishers.

U.S. EPA. 1977. Water Supply Research, Office of Research and Development. *Ozone, Chlorine Dioxide and Chloramines as Alternatives to Chlorine for Disinfection of Drinking Water.* Cincinnati, Ohio: U.S. EPA.

U.S. EPA. 1993. *Guidance Manual for Compliance with the Filtration and Disinfection Requirements for Public Water Systems Using Surface Water Sources.* Science and Technology Branch, Criterion and Standards Division, Office of Drinking Water, U.S. EPA.

Yoo, R. C., D. R. Brown, R. J. Pardini, and G. D. Bentson. 1995. "Microfiltration a Case Study." *Journal AWWA* 87(3): 38–49.

INDEX